Using Multimedia Tools and Applications on the Internet

DENNIS O. GEHRIS

Bloomsburg University

An Imprint of Wadsworth Publishing Company

I(T)P® An International Thomson Publishing Company

Belmont, CA • Albany, NY • Bonn • Boston • Cincinnati • Detroit • Johannesburg • London
Madrid • Melbourne •Mexico City • New York • Paris
Singapore • Tokyo • Toronto • Washington

New Media Publisher: Kathy Shields
Assistant Editor: Shannon McArdle
Production: Vicki Moran/Publishing Support Services
Composition: Publishing Support Services
Copyeditor: Victoria Nelson
Print Buyer: Karen Hunt
Permissions Editor: Robert Kauser
Cover Design: Ark Stein/The Visual Group
Printer: Edward Brothers

Printed in the United States of America
1 2 3 4 5 6 7 8 9 10

For more information, contact Wadsworth Publishing Company, 10 Davis Drive, Belmont, CA 94002, or electronically at http://www.thomson.com/wadsworth.html

International Thomson Publishing Europe
Berkshire House 168-173
High Holborn
London, WC1V 7AA, England

Thomas Nelson Australia
102 Dodds Street
South Melbourne 3205
Victoria, Australia

Nelson Canada
1120 Birchmount Road
Scarborough, Ontario
Canada M1K 5G4

International Thomson Publishing GmbH
Königswinterer Strasse 418
53227 Bonn, Germany

International Thomson Editores
Campos Eliseos 385, Piso 7
Col. Polanco
11560 México D.F., México

International Thomson Publishing Asia
221 Henderson Road
#05-10 Henderson Building
Singapore 0315

International Thomson Publishing Japan
Hirakawacho Kyowa Building, 3F
2-2-1 Hirakawacho
Chiyoda-ku, Tokyo 102, Japan

International Thomson Publishing Southern Africa
Building 18, Constantia Park
240 Old Pretoria Road
Halfway House 1685, South Africa

Library of Congress Cataloging-in-Publication Data
Gehris, Dennis.
Using multimedia tools and applications on the Internet / Dennis O. Gehris.
p. cm.
Includes index.
ISBN 0-534-51939-3
1. Multimedia systems. 2. Internet (Computer network) I. Title.
QA76.575.G48 1998 96-49781

To Deborah
Without her encouragement and support,
this book could not have been written.

CONTENTS

CHAPTER 6 *Video and Animation on the Internet 121*

CHAPTER 7 *Computer Conferencing on the Internet 139*

CHAPTER 8 *Videoconferencing on the Internet 163*

PREFACE

Until recently, multimedia concepts and principles and the Internet were taught separately. Multimedia computer applications required separate computer hardware and software to create the desired graphics, sound, and video because only UNIX text-based applications such as telnet, gopher, and file transfer protocol (FTP) were available for the Internet. Today, with the advent of the World Wide Web (WWW), we have seen a marriage of multimedia and the Internet. This union has provided a great deal of excitement among academic and nonacademic communities alike, creating an unprecedented interest in the Internet.

This book explains multimedia and the Internet and tells you how to put the two together. It focuses on the various multimedia tools and applications on the Internet available either free or as inexpensive shareware for the IBM or IBM PC–compatible and Apple Macintosh microcomputers. The book serves as a guide for the most common multimedia tools and applications relating to hypermedia, graphics, audio, video, animation, computer conferencing, videoconferencing, Adobe Acrobat, ASAP WebShow, Macromedia Shockwave, and virtual reality.

The text includes a detailed explanation of each of the multimedia tools and applications along with step-by-step instructions for using the most popular software in each of the areas. Computer screen visuals (screen dumps) are liberally used to illustrate the steps. Questions at the end of each chapter are included for users to test their knowledge of the material, and exercises provide an opportunity to apply the concepts. Most exercises require a microcomputer with a connection to the Internet.

The majority of the examples apply to the Windows 3.1 and Windows 95 platforms. Some examples are also provided for the Apple Macintosh computer.

MARKET

The market for this textbook includes the following course areas: educational technology, information technology, instructional media, software applications, introduction to computer science, telecommunications, business communication, or any other course in which the Internet or multimedia is taught. Additional markets for the book outside higher education are community education, corporate training, business development courses, and home computer enthusiasts.

UNIFORM RESOURCE LOCATORS

The uniform resource locators (URLs) in this textbook were correct at the time that it was written. However, keep in mind that URL addresses may change or disappear without notice. This is one of the wonders and frustrations of the Internet. Sometimes a new URL will be provided on the old page, but often you will receive error messages indicating that the server does not have a DNS entry or will receive a "server not found" message. There are several options available if you receive this message. One option is to use one of the search engines discussed in Chapter 2 to find the new URL. A second thing that you can do is to link to the Integrated Media Group Resource Center Web page at Wadsworth Publishing Company (http://www.thomson.com/rcenters/img/img.html). Find the link for this textbook and check the User's Guide to determine if a revised link is available.

ONLINE GUIDES

A User's Guide and an Instructor's Manual for this text are available online. Related information for both students and instructors can be found at the Wadsworth Publishing Company Integrated Media Group Resource Center Web page. The URL for this page is http://www.thomson.com/rcenters/img/img.html.

User's Guide The online User's Guide provides the following:

- The purpose of each chapter and the learning goals
- Links to the uniform resource locators (URLs) mentioned in this textbook with updates on revised links. Users of this text will be able to link to the sites by clicking on the hypertext. (See Chapter 2 for an explanation of URLs.)
- Links to the URLs in the end-of-chapter exercises

Instructor's Manual The online Instructor's Manual provides the following:

- General teaching suggestions
- Teaching suggestions and comments for each chapter and appendix
- Additional resources

ACKNOWLEDGMENTS

The author wishes to express gratitude to the following colleagues for their thoughtful critiques of the text: Meral Binbasioglu, Hofstra University; Les Blackwell, Western Washington University; Burt Geeene, Purchase College–State University of New York; Virginia Plumley, Marshall University; Erik Rolland, University of California, Riverside; Mark A. Rosso, Meredith College; LeRoy J. Tuscher, Lehigh University; Emery D. Twoey, Olivet Nazarene University; and Vicki Wise, The Definance College.

Dennis O. Gehris, Ed.D.
Bloomsburg, Pennsylvania

CHAPTER

1

Introduction to the Internet and Multimedia

This chapter introduces you to the Internet and multimedia and lays the groundwork for understanding multimedia applications as an integral part of the Internet.

What You Will Learn

- How the Internet is defined
- How to access the Internet
- A brief history of the Internet
- Ways to use the Internet
- How multimedia is defined
- Multimedia levels
- Multimedia hardware requirements

The Internet Defined

The Internet has recently received a great deal of publicity, and everyone seems to want an Internet connection. But there is still widespread misunderstanding about the Internet, its capabilities, and its uses. What *is* the Internet? The **Internet** is an interconnected network of computer systems encompassing about 130 countries. All the connected computer systems share one common protocol: **transmission control protocol/Internet protocol (TCP/IP)**. A **protocol** is the set of rules or conventions by which two machines talk to each other. The TCP/IP allows computer systems of various types to communicate with each other within the Internet.

Internet Access

Access to the Internet is available in the following ways: by an electronic mail connection, a direct connection, an ISDN connection, a shell connection, a SLIP/PPP connection, or a commercial online service. The least expensive way to obtain Internet access is by working for a company with an Internet connection or by becoming a student. Most colleges and universities, including community colleges, now provide Internet access to all enrolled students. In addition, some public libraries have Internet terminals for anyone to use, and many communities offer connections to a service called a FreeNet that you can use from a library or school or by dialing in.

ELECTRONIC MAIL CONNECTION

An *electronic mail (e-mail) connection*, usually provided by a company or other organization, limits users to sending mail electronically outside the organization via the Internet. Other Internet applications are generally not available through e-mail, so it would not be a good choice for using multimedia on the Internet.

DIRECT CONNECTION

A *direct (or dedicated) connection*, currently the preferred way of accessing the Internet, is made through an individual's company, educational institution, or other organization. Some organizations provide dial-up access via a modem. A direct connection is usually accomplished through a T1 transmission line, which offers a transmission speed of 1.544 megabytes (MB), as compared with a voice transmission line's 28.8 kilobytes (KB).

ISDN CONNECTION

The *integrated services network (ISDN) connection* is also designed to carry large amounts of information at a fast rate of speed. The ISDN is especially well suited for transmission of high-quality audio (see Chapter 5) and video (see Chapter 6) as well as desktop videoconferencing (see Chapter 8) for users without a direct connection. The regional telephone companies in the United States are gradually making ISDN available in more localities. Currently, there is a better than 75 percent coverage, and this percentage will be increasing in the near future. Most long-distance ISDN connections within the United States are transmitted between 56 KB and 128 KB.

SHELL OR SLIP/PPP CONNECTIONS

Private providers either offer a shell connection or a serial line Internet protocol/point-to-point protocol (SLIP/PPP) connection to the Internet. A *shell connection* is a dial-up service in which you access your account with terminal emulation software. The Internet software runs on the provider's host computer and the emulation software displays the link between the local host and the Internet computers. With a SLIP/PPP connection, your computer becomes physically connected to the Internet. The connection is made through a private provider using dial-up, but special software allows TCP/IP packets to be transmitted between the Internet-connected computer and your computer. This is the second best connection after a direct or ISDN connection. Subscribers to these Internet providers usually pay an installation fee and a per-month fee that will either provide them with an unlimited number of access hours or will limit them to a specific number of hours before additional charges are added.

COMMERCIAL ONLINE SERVICE CONNECTION

The last way to gain Internet access is through commercial online services such as America Online, CompuServe, and Prodigy. The number of Internet services these providers offer varies, but all have electronic mail transmission to other individuals connected to the Internet. Many of the services have recently provided subscribers access to Internet servers in multimedia format. Subscribers are usually charged a monthly fee for a specific number of hours of basic services. Special services often result in additional fees above the basic fee. Figure 1-1 shows a Windows CompuServe Information Manager screen listing the Internet services available.

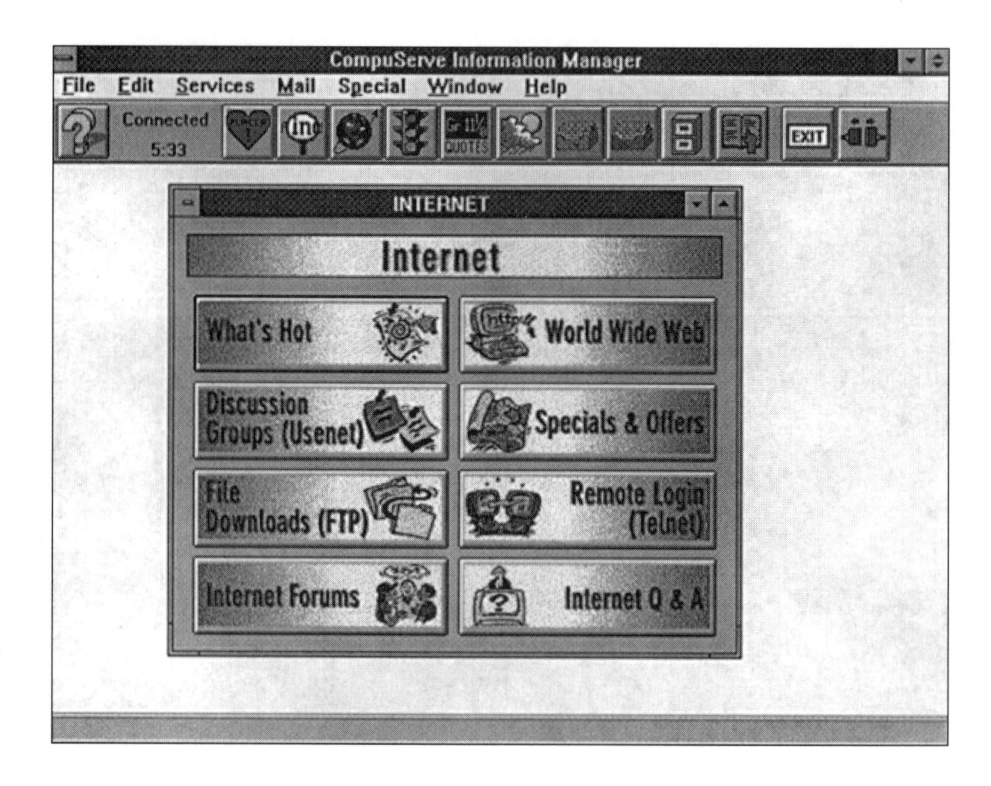

Figure 1-1. Internet Services on the CompuServe Information Manager

Brief History of the Internet

How did the Internet develop into what it is today? As you will see, it went through a number of transformations. The Internet began in 1969 as the Advanced Research Projects Agency Network (ARPANET), a project of the U.S. Department of Defense. The Department of Defense made ARPANET available to universities and other organizations. Research was then conducted to determine if a network could be developed that would ensure that military communications could continue in the event of a war. In response to this challenge, ARPANET became a *packet-switched* network in which information gets broken into little packets that move independently of each other until they reach their destination.

In 1983, the system continued to expand and split into two networks—ARPANET and MILNET. ARPANET was used for civilian research efforts and MILNET was reserved for military use. Several other new networks were also formed during the early 1980s to serve other groups and organizations. One of these was the Because It's Time Network (BITNET) for academic communities. Another, the Computer Science Network (NSNET), connected researchers together for the sharing of information. In 1986, the National Science Foundation connected researchers across the country with five supercomputer centers that became known as NSFNET. This network formed a backbone of transmission lines—fiber-optic wires, satellite links, and microwaves—to carry large amounts of traffic very quickly over long distances. This NSFNET backbone became the basic infrastructure for the Internet.

Internet Uses

The Internet today has four main uses: electronic mail, remote log on, file transfer, and multimedia. **Electronic mail** (e-mail), the most used application, permits individuals with an Internet connection who have been assigned a user identification (i.e., jsmith@psu.edu) to send text messages to others connected to the Internet. **Remote log on** (or telnet) allows individuals using a computer with an Internet connection to log on to any other computer connected to the Internet. They can, for example, check for e-mail messages and perform other tasks. **File transfer** is the ability to transfer a file from one computer to another on the Internet. **File transfer protocol** (ftp) is one utility that permits file transfer. Others are archie and veronica, which provide the means for searching the Internet for information on specific topics.

The fourth and newest feature of the Internet is multimedia. Before this application became available, it was only possible to view and communicate information in simple text format, as shown in Figure 1-2. In the past, when the Internet was used mainly by government and academia for research purposes, the nongraphical interface was adequate. Now multimedia makes it possible to view and communicate information in a range of forms that will be discussed in the next section.

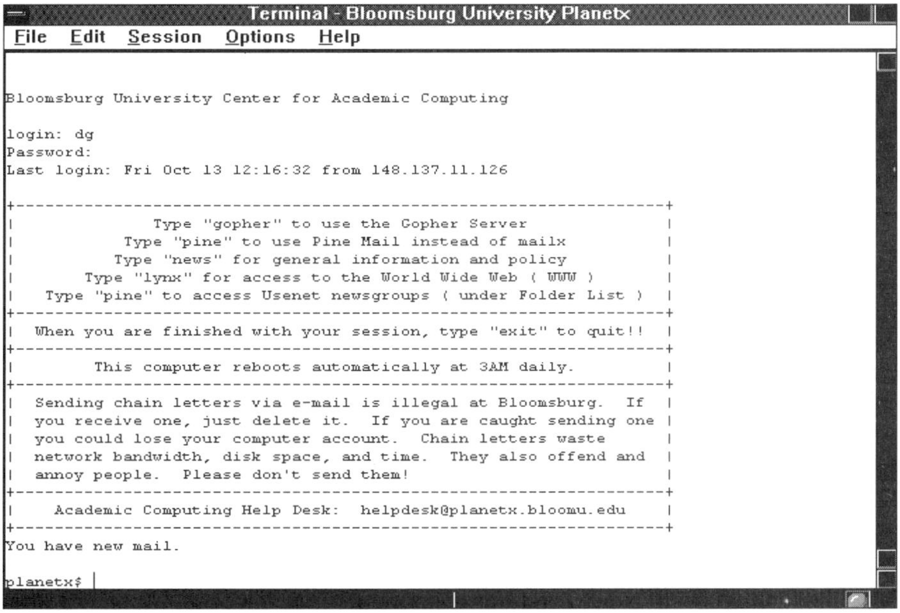

Figure 1-2. Simple Text Format

```
Terminal - Bloomsburg University Planetx
File  Edit  Session  Options  Help

Bloomsburg University Center for Academic Computing

login: dg
Password:
Last login: Fri Oct 13 12:16:32 from 148.137.11.126

+------------------------------------------------------------------+
|               Type "gopher" to use the Gopher Server              |
|            Type "pine" to use Pine Mail instead of mailx          |
|           Type "news" for general information and policy          |
|        Type "lynx" for access to the World Wide Web ( WWW )       |
|    Type "pine" to access Usenet newsgroups ( under Folder List )  |
+------------------------------------------------------------------+
|  When you are finished with your session, type "exit" to quit!!  |
+------------------------------------------------------------------+
|         This computer reboots automatically at 3AM daily.         |
+------------------------------------------------------------------+
|  Sending chain letters via e-mail is illegal at Bloomsburg.  If   |
|  you receive one, just delete it.  If you are caught sending one  |
|  you could lose your computer account.  Chain letters waste       |
|  network bandwidth, disk space, and time.  They also offend and   |
|  annoy people.  Please don't send them!                           |
+------------------------------------------------------------------+
|    Academic Computing Help Desk:  helpdesk@planetx.bloomu.edu     |
+------------------------------------------------------------------+
You have new mail.

planetx$
```

Multimedia Defined

Many people have heard the term *multimedia* but don't really know what it means. This is probably because the term has been connected with many technological tools. Multimedia means different things to different people. Some associate multimedia with video games, others with voice-activated devices.

Multi means more than one, and *media* means a form of communication. **Multimedia** on the Internet means the integration of at least two media. These media can include text, photos, graphics, sound, music, animation, and full-motion video. Think of multimedia on the Internet as a combination of technologies. A technology is something that makes something else more efficient, and the use of multimedia makes computers more efficient and easier to use by providing them with more capabilities.

The Internet provides multimedia to users through the **World Wide Web** (**WWW**, or **Web**), a global hypermedia system. The World Wide Web not only provides information in multimedia format, it also provides the information interactively, sending and receiving input from users via their mouse or keyboard. Interactivity is made possible through **hypermedia**, in which users click on text or other objects to link to other objects, text, pages, or site on a World Wide Web page. When text provides the link, it is called **hypertext**. Hypermedia, however, is a more general term that includes both textual and nontextual features.

Figure 1-3. Multimedia Components

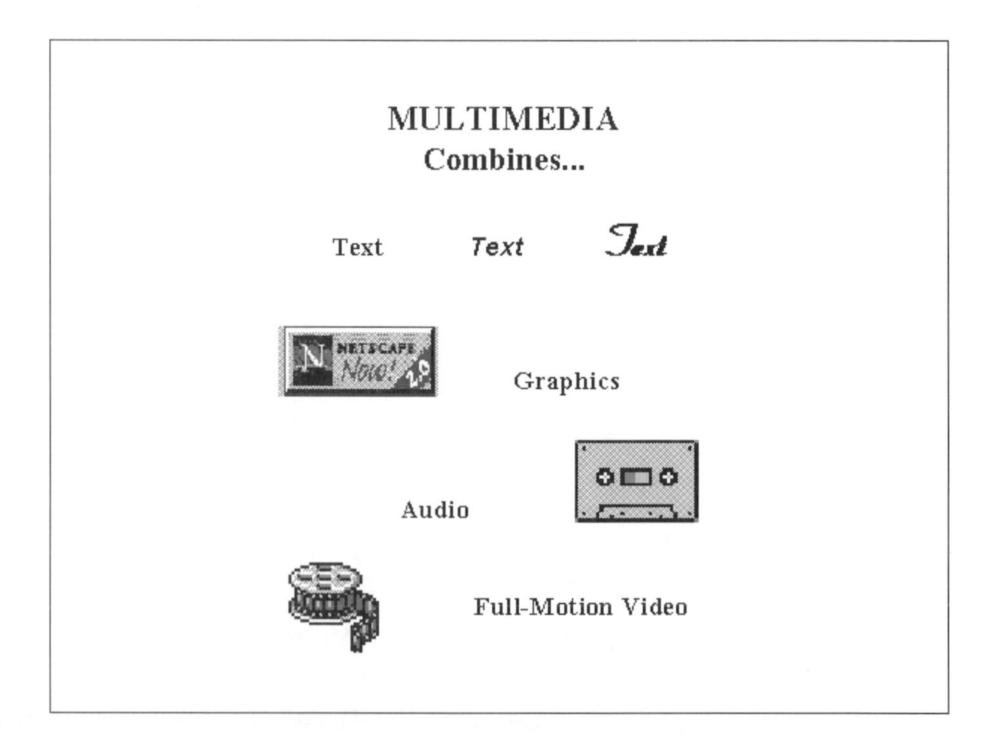

Multimedia Levels

Not all multimedia is interactive. Seven levels of multimedia are identified in Table 1-1, and only two of them are interactive. Levels 1 and 2 utilize black-and-white or color text and graphics and are displayed onscreen as static—that is, there is no movement associated with these elements. Software that helps produce multimedia at these two levels includes Delta Graph Pro, Freelance Graphics, Harvard Graphics, Impact, Persuasion, PowerPoint, Super Show & Tell, and WordPerfect Presentations.

Multimedia levels 3, 4, and 5 utilize some form of video and animation. Software that helps produce multimedia at these three levels includes the software listed for levels 1 and 2 along with Action, Astound, Charisma, Compel, MovieWorks, Special Delivery, and Vmedia. Additional support programs such as Adobe Photoshop, Illustrator, Premiere, Fractional Painter, or Strata Studio Pro can also be used.

Multimedia levels 6 and 7 encompass some form of interactivity, either through group or individual interaction. Software that helps produce multimedia at these levels includes Apple Media Kit, Authority Multimedia, Authorware Professional, Digital Chisel, Everest Authorizing Systems, Guide Author, Course Builder, Image Q, Macromind Director, Multimedia Toolbook, and Ultimedia Builder/2. A graphical Web browser would belong on level 7 since it provides input through individual interaction via the hypermedia links.

TABLE 1-1 MULTIMEDIA LEVELS

Type	Level	Elements
Static	1	Black-and-white text and graphics
	2	Color text and graphics
Animated	3	Simple animated text and graphics
	4	Predigitized video and animation
	5	Authored video and animation
Interactivity	6	Input required from the audience on a group basis
	7	Input required from the audience through individual interaction

Hardware Requirements for Multimedia

Will any computer hardware configuration be adequate for multimedia? The answer, unfortunately, is no. You need to have specific types of hardware for both IBM and IBM PC–compatible and Macintosh computers. Equipment is often purchased with standard configurations for nonmultimedia applications such as word processing, spreadsheeting, and similar applications. Today some

PCs are being sold as "multimedia computers" or "multimedia capable." Is this label a guarantee that they will work well with all of your multimedia applications? The answer is, not necessarily. Keep reading to learn what features to look for in purchasing or setting up a computer for multimedia.

IBM AND IBM PC-COMPATIBLE COMPUTERS

Although IBM and IBM PC–compatible computers were once not as well suited as the Macintosh for work with multimedia, this is no longer the case. When you purchase IBM or IBM PC–compatible computer hardware for multimedia, you must consider the following components: monitors and video cards, central processing units (CPUs) and motherboards, hard drives, CD-ROM drives, sound cards, and other items (see Table 1-2). This section will briefly describe each of these components and will make recommendations for each type in configuring multimedia with the Internet.

TABLE 1-2 IBM AND IBM PC-COMPATIBLE HARDWARE COMPONENTS TO CONSIDER FOR MULTIMEDIA ON THE INTERNET

Monitor and Video Card
CPU and Motherboard
Hard Drive
CD-ROM Drive
Sound Card
Other Items

Monitor and Video Card You probably already know the purpose of a monitor. Together with the video card, it is the main output device for a computer. These two pieces of hardware work together to determine how well you can view the various visual elements of multimedia—text, graphics, and video. There are several points to consider when you purchase a monitor and video card: mode and resolution, size of the monitor, dot pitch, whether the monitor is interlaced or noninterlaced, the refresh rate, whether it meets Swedish Low Emission Standards, driver software, and local bus capabilities. (A bus is an internal electrical pathway along which signals are sent from one part of the computer to another.)

It makes sense to spend the extra dollars on a monitor. A motherboard or CPU will probably be out of date in a short period of time, but a good monitor will last for many years. Get the largest, noninterlaced, low-emission, antiglare, flat screen, Super VGA monitor with the smallest dot pitch you can afford. Purchase a local bus controller card, because multimedia applications on the Internet are video intensive. You should ask to see the monitor with your video card before purchasing your system, since some monitor and video card combinations cause flicker or blurry images on the screen. You definitely don't want to experience problems in viewing your text, graphics, or video.

CPU and Motherboard The central processing unit (CPU) and the motherboard are the heart of a multimedia computer system. The CPU is a computer's internal storage, processing, and control circuitry. The motherboard is a large circuit board that contains the CPU and other components. A computer with a powerful CPU and sufficient memory will execute any program more quickly. Nothing is more annoying than a computer with slow response time when you are executing a multimedia helper application using a World Wide Web browser program on the Internet.

In purchasing a motherboard and CPU, you need to consider type of bus, whether it has a local bus, amount of random access memory (RAM), BIOS, type of CPU, and CPU speed. For multimedia applications on the Internet, a Pentium 75-megahertz (MHz)–based computer or better should be purchased. (One MHz equals 1 million electrical vibrations or cycles per second.) Although a 486 system is adequate, it is better to purchase the most up-to-date technology. Most hardware manufacturers have now ceased production of the 486 systems. As far as RAM is concerned, 16 MG should be the minimum, especially with Windows 95. The motherboard should have the capacity to expand to 32 MB or more and include VLB or PCI local bus technology.

Hard Drive The hard drive is an important component of a multimedia computer system because most of these programs are large and require a substantial amount of space on a hard drive. Even CD-ROM programs need several megabytes of storage space. In purchasing a hard drive, consider its size, type, control, and access time. The storage capacity of a hard drive is measured in either megabytes (MB) or gigabytes (GB). (A MB is approximately 1 million bytes; a GB is about 1 billion bytes.) Hard drives either use RLL, MFM, IDE, or SCSI (pronounced "scuzzy") technology; however, integrated drive electronics (IDE) and small computer system interface (SCSI) are the types of drives in use today. SCSI drives are larger, faster, and more expensive than IDE drives.

If money is not a consideration, get a SCSI-2 hard drive and controller card, which will chain up to seven devices. If money *is* a consideration, IDE hard drives on a VLB controller card will perform well. Many new CD-ROM drives require an IDE controller, so combining an IDE hard drive and an IDE CD-ROM drive makes sense. SCSI drives are appropriate if the system needs a huge amount of hard drive space or if chained devices or a SCSI CD-ROM is part of the computer system. For multimedia applications on the Internet, IDE hard drives and controller cards are a good choice for most systems. A size of 1 GB or better is desirable.

CD-ROM Drive CD-ROM drives allow the computer to read programs and data recorded on a compact disk (CD). A CD can hold as much as 650 MB of a mixture of program, audio, and video data or 74 minutes of digital audio data. This is the equivalent to about 464 high-density floppy diskettes. The CD-ROM drives are write once read many (WORM) devices. This means that the data are recorded on them one time and then can be read many times by the user. CDs seem to be the delivery medium that many software manufacturers have chosen because they are inexpensive to duplicate after the master is created.

In purchasing a CD-ROM drive, consider speed, interface, compatibilities, software, cables, and caddies. CD-ROM drives are, in ascending order of access speed, single-spin, dual-spin, triple-spin, quad-spin, eight-spin, and ten-spin. (The lower-speed drives may no longer be available for purchase.) They also require a controller, which can be proprietary, SCSI, SCSI-2, or IDE. CD-ROM drives require the disk either to be placed directly in the CD-ROM drive or placed first in a CD caddy before it is inserted in the drive.

Is a CD-ROM drive absolutely necessary when you are working on the Internet? Using multimedia applications on the Internet probably does not require a CD-ROM drive to access the World Wide Web with a browser program. However, some helper application programs such as Microsoft Video for Windows, which you will need to use CU-SeeMe videoconferencing, comes packaged on a CD. Also, some excellent instructional programs pertaining to multimedia and the Internet are CD-based programs.

If you are purchasing a new computer, then you may want to consider purchasing a system with a CD-ROM drive. Be sure, however, that the system contains a double-speed drive or better. If you plan to play full-motion video from the CD-ROM drive, a quad-speed or better drive is recommended. If you would like to add a CD-ROM and a sound card to an existing computer, a CD-ROM/Sound Card package deal is a good idea; this way all the parts are included and are compatible.

Sound Card Although it is possible to play sounds through a computer's built-in speaker, a *sound card* will allow the computer to produce better-quality recorded music, voices, and other sounds. Sound cards use two methods to create sounds: FM Synthesis and WaveTable. A card using FM Synthesis generates sound synthetically; the WaveTable card uses digitized sound. Most sound cards sold today are 16-bit and produce better sound than the older 8-bit cards. The sound card needs a line-level input and/or a microphone input to be able to record audio or to send audio over the Internet. A Sound Blaster–compatible 16-bit sound card is a good choice, but if you expect to use CU-SeeMe videoconferencing, then a full duplex sound card that sends and receives audio simultaneously will work best. The following sound cards are full duplex and are known to work with CU-SeeMe: Pro Audio Spectrum, Gravis Ultrasound, Turtle Beach Tropex, and Logitech MovieMan System.

Other Items You may want to consider purchasing several other optional items for using multimedia on the Internet. You will need a *video capture card* and a *video camera* if you plan to use CU-SeeMe videoconferencing (see Chapter 8), to video-capture still pictures or to record video clips for a World Wide Web home page. You may need a *microphone* to record sounds, although some sound cards and upgrade kits come bundled with a microphone, and a *scanner* if you want to digitize still pictures and save them as files to be used on a home page. You'll need an LCD panel, a large screen monitor, or a projection unit to display a multimedia session using a World Wide Web browser program to a large group. Finally, you need a modem for dial-up connections to the Internet. Table 1-3 summarizes hardware recommendations for multimedia applications on the Internet.

TABLE 1-3 RECOMMENDATIONS FOR IBM AND IBM PC-COMPATIBLE HARDWARE COMPONENTS FOR MULTIMEDIA ON THE INTERNET

Component	Recommendation
Monitor and Video Card	The largest, noninterlaced, low-emission, antiglare, flat screen, Super VGA monitor with the smallest dot pitch that you can afford.
CPU and Motherboard	Pentium-75 MHz or better; 16 MG of RAM or more with capacity to expand to 32 MB or more; VLB or PCI local bus technology
Hard Drives	SCSI-2 hard drive and controller card or an IDE hard drive on a VLB controller card; 1 GB or better
CD-ROM Drive	Double-speed or better
Sound Card	Sound Blaster–compatible 16-bit; full-duplexed for CU-SeeMe
Other Items (optional)	Video capture card and a video camera; microphone; scanner; LCD panel, a large screen monitor, or a projection unit; modem

MACINTOSH COMPUTERS

Because Macintosh computers come preequipped with many multimedia capabilities, you need to consider fewer peripherals than with IBM or IBM PC–compatible computers. Although exact specifications may vary, here are some specific recommendations for using a Macintosh with multimedia applications (see Table 1-4).

CPU, Motherboard, and System Software A Power Macintosh, 100 MHz or higher, is recommended for multimedia. In addition, 16 MG of RAM or more and 16-bit stereo sound is needed. System 7 software or higher is suggested.

Monitor A 14-inch or larger, .28-mm or more dot pitch, RGB color monitor is recommended.

Hard Drive The recommended size is 1 GB.

CD-ROM Drive A double-speed drive or better is recommended.

Other Items All of the items previously mentioned for the IBM or IBM PC–compatible computer, with the exception of a video capture card, are also desirable for the Macintosh. These include video camera, microphone, scanner, and an LCD panel.

TABLE 1-4 RECOMMENDATIONS FOR MACINTOSH HARDWARE COMPONENTS FOR MULTIMEDIA ON THE INTERNET

Component	Recommendation
Processor	Power Macintosh, 100 MHz or more with 16-bit stereo sound
Memory	16 MG of RAM or more
System Software	System 7 or higher
Monitor	14-inch or larger, .28-mm dot pitch or more RGB color
Hard Drive	1 GB
CD-ROM Drive	Double-speed or better

Questions for Review

1. What is the Internet?
2. What are the ways in which the Internet can be accessed?
3. List three significant events in the history of the Internet.
4. What are the uses of the Internet?
5. What is multimedia?
6. How many multimedia levels are available on the Internet, and what is an example of each?
7. List the multimedia hardware requirements for an IBM or IBM PC–compatible computer in each of the following areas: monitor and video card, CPU and motherboard, hard disk, CD-ROM drive, sound card, and other features.
8. List the hardware requirements for a Macintosh computer for multimedia: CPU, motherboard, and system software; monitor; hard drive; and CD-ROM drive.

Exercises

If you are using an IBM or IBM PC–compatible computer, use file transfer protocol (ftp) or the Macintosh Fetch utility to complete the following exercises. See Appendix A for information on downloading files.

1. Download and print the file internet.faq using one of the following hosts/paths:

   risc.ua.edu/pub/network/pegasus/FAQs
   ftp.ci.cuslm.ca/pub/pegasus/FAQs

   Read the article and answer the following questions:
   a. What are the steps in the process of deciding what kind of access you want to the Internet?
   b. What is WinSock?
   c. What is the simple mail transfer protocol (SMTP)?
   d. What is post office protocol (POP)?

2. Download and print the file answers.to.new.user.questions using the following host/path:

   nic.merit.edu/introducing.the.internet

   Read the article and answer the following questions:
   a. What is the difference between the Internet and an internet?
   b. What is an advantage to the domain name system (DNS)?

3. Download and print the file netiquette.txt using one of the following hosts/paths:

   thedon.cac.psu.edu/pub/people/dlp/TRDEV-L
   ftp.sunet.se/pub/Internet-documents/doc/netiquette

   Read the article and answer the following questions:
   a. What is netiquette?
   b. Why is it important to have rules of conduct for using the Internet?

CHAPTER 2

The World Wide Web (WWW)

This chapter presents the World Wide Web on the Internet as an important multimedia application. It explains how the Web operates, the various types of Web browsers, how to use Netscape (the most popular graphical browser in use today), and how to conduct a search on the Web.

What You Will Learn

- How the World Wide Web is defined
- How information is accessed on the World Wide Web
- Features of World Wide Web browsers
- Types of World Wide Web browsers
- Steps and procedures for using Netscape
- How to search on the World Wide Web

The World Wide Web Defined

The **World Wide Web** (**WWW** or Web) is what makes multimedia applications possible on the Internet. Officially, it is defined as a "wide-area hypermedia information retrieval initiative designed to give universal access to a large universe of documents." The WWW started in 1989 at CERN, the European Laboratory for Particle Physics in Geneva, Switzerland and was funded by eighteen European member states. CERN researchers developed a system that would allow physicists and other scientists to share information with others and to provide for information in textual form based on *hypertext transfer protocol* (*http*). There was no intention of adding video or sound, and the capability of transmitting images was not considered. Once the CERN researchers established the http specification, client and server software was written and the WWW was on its way to becoming a reality.

The first piece of WWW software had the capability to view and transmit hypertext documents to users on the Internet and to edit hypertext documents onscreen. Demonstrations were given to CERN committees and seminars as well as to the Hypertext conference in 1991. As the WWW developed, the new software included capabilities of displaying enhanced text, graphics, and other multimedia components.

The advantages of the WWW become apparent to anyone who has used any of the other Internet services, such as file transfer protocol (ftp), gopher, and the wide area information service (WAIS). Using the WWW, less time is spent in learning procedures for obtaining information on the Internet and more time on actually obtaining the information. Gopher, for example, which works with a system of menu items, can be very inefficient and time-consuming; menu items simply may not represent the type of information desired, and users often end up aimlessly wandering through menus. These problems can be avoided on the WWW, whose pages also use lists but usually include text that describes the information.

The main difference between the WWW and the text-based portion of the Internet is that the former uses **hypermedia**. As we saw in Chapter 1, hypermedia is made up of hypertext and other multimedia elements that allow users to move from item to item and back again without having to follow a predefined path. The term *hypertext* was coined in the 1960s by Ted Nelson to describe a means of moving through text in a nonlinear manner. These nonlinear paths, called **links**, are connected to other resources on the Internet. Links can be part of a sentence or paragraph or may be connected to an image or digitized sound. When the link is in the form of hypertext, the text is often underlined with a graphical browser. Because it is underlined, the user knows that clicking on the word or words with the mouse cursor results in a connection to a predefined place on the Internet. When a link is connected to an image, text will usually instruct the user to click on the image for a connection. There is also a way of reconnecting with previous links, so that the user can return to former sites as desired.

WWW pages are developed using the **hypertext markup language** (**HTML**), which provides a system for marking up text documents and a way to integrate multimedia and the use of hyperlinks. Although not a difficult language to use, HTML uses special tags through a specific structure that users must learn. Chapter 3 will provide a more detailed explanation of HTML.

Who currently uses the World Wide Web? Research in 1995 and 1996 indicates the following characteristics of World Wide Web users:

- Total number: 55,000,000
- Growth rate: the number of users doubles every ten weeks
- 73 percent are male; 27 percent female
- 13 percent are under the age of 20
- 37 percent are between 21 and 35 years old
- 45 percent are between 36 and 59 years old
- 3 percent are older than 60 years old

In the past, most WWW users were probably students at four-year college campuses within the United States. Although it is difficult to know who is using the WWW at the present time, it is probable that in the future many more individuals will have the opportunity to experience the power of the WWW. This is due partly to the WWW's increased availability through online services (America OnLine, CompuServe, Prodigy, etc.). Through these services, individuals will be able to experience the value of retrieving information in a variety of multimedia formats.

How is the WWW used for multimedia applications? Consider the following examples:

- Jim Durane reads a text on the Japanese language. When he selects a Japanese phrase, he hears the phrase spoken in that tongue.
- Nina Smithson is a law student studying the New York statutes. By selecting a passage, she finds precedents from a 1932 U.S. Supreme Court ruling stored at Yale University. Cross-referenced hyperlinks allow her to view any one of 535 related cases with audio annotations.
- George Johannson is a scientist doing work on cooling steel springs. By selecting text in a research paper on the Internet, he is able to view a computer-generated movie of a cooling spring.
- Susan Schaffer is a student reading a digital version of an art magazine. She is able to select a work to print or display in full and can rotate movies of sculptures.
- Bob Brown is a sales manager who needs to find new ways to promote his company's products. He is able to develop a Web site that provides information and graphics on products as well as a way to place orders.

Accessing Information on the World Wide Web

Each of the links defined in hypermedia documents on the World Wide Web represents a **uniform resource locator** (**URL**). You need to know the URLs for the Web sites you want to access if a link to a location on the Internet doesn't happen to be on the WWW page on which you happen to be working. URLs are often cited in articles and advertisements and represent a starting point for linking to other URLs through hypermedia. WWW users often use a URL as a starting point that is known as a **homepage**. The basic form of a URL is service://domain name/full path name.

TYPES OF URLS

The following types of uniform resource locators (URLs) are available: http, ftp, gopher, and telnet. Not all of these types of URLs may be available on all browsers.

http The service **hypertext transport protocol** (**http**) is the Internet standard that supports the exchange of information on the World Wide Web. It is used by Web clients and servers to exchange documents in the following way: a client opens a connection to a server, the client requests a particular document, the server answers the request, and the connection is closed. An example of a URL of this type would be http://www.yahoo.com for the Yahoo list of WWW resources.

ftp We mentioned ftp in Chapter 1. Let's define it more comprehensively. The service **file transfer protocol** (**ftp**) is used to exchange files on the Internet. An ftp URL usually indicates that the file can be retrieved by anonymous ftp so that a password is not required. An example of an ftp URL is ftp://sumex-aim.stanford.edu/info-mac/comm/info/ftp-primer.txt, a file that provides information about anonymous ftp.

gopher The name refers to a furry animal that lives in the ground, but it also represents another Internet URL service. **Gopher** servers can be accessed through URLs that indicate the path to these servers. Each gopher connection results in a different list of *menu* items that is represented on a browser by a series of labeled folders. Click on the folders to use the various services or to access *submenus*, which are also represented by a folder list. An example would be gopher://marvel.loc.gov, which would connect you to the gopher server at the Library of Congress.

telnet **Telnet** allows you to log on to a remote computer connected to the Internet to access information or to run programs. Not all browsers permit you to use telnet, and some browsers require that you configure a **helper application program** on your hard disk for telnet. If you don't already have a telnet application that allows you to use telnet from within the browser, you can download one. A telnet application for Windows-based computers can be found at ftp://gatekeeper.dec.com/pub/micro/msdos/win3/winsock/trmptel.zip or, for Macintosh computers, ftp://ftp.utexas.edu/pub/mac/tcpip/ncsa-telnet-26.hqx.

An example of a telnet URL is telnet://locis.loc.gov, which is another way to connect to the Library of Congress.

Features of World Wide Web Browsers

What type of software do you need to use the World Wide Web? Once you have an Internet connection, a **WWW** (or **Web**) **browser** (also known as a **Web client**) is the software program needed to make access to Web servers on the Internet possible. All Web browsers have some basic features. They must

- Have an easy, simple installation procedure. Fortunately, the latest versions of most Web browsers possess this feature.
- Send requests for data to Web servers in the correct Internet and World Wide Web formats. This very necessary feature is incorporated into all Web browsers.
- Receive and display information that is sent back from Web servers or display error messages to the user. This is also a necessary feature that all Web browsers possess.

In addition, several other browser features are desirable, though not all browsers possess all of them. (A browser with most of the features might work well for the inexperienced user.) They should

- Display information in the original design format of the document's author. This is not always possible, especially with a nongraphical browser and some graphical browsers.
- Save documents that arrive via requests made by the user. This feature allows you to view the text and layout information for homepages at a later time without having to connect to them again.
- Store bookmarks or hotlist entries of locations that the user might want to revisit. This feature is a real timesaver because it allows the user to link quickly to pages accessed during previous sessions.
- Maintain a history record of links, often done by using a *cache*, saved on the user's hard disk. This feature provides a means of revisiting previous links without having to download them again during the same session.

- Turn on and off **inline images**, graphic images that are transmitted over the Web. This is a valuable feature for those individuals who are using a dial-up connection to the Internet because it greatly improves the speed at which the information is received.
- Print pages. There are times when you might want to have a hard copy of a Web page, especially if it contains information you need for research.
- Copy information from a homepage to the Windows Clipboard. This feature provides the ability to clip information from the browser program to another Windows program (such as a word processing program).
- Display the first screen of text while the graphical elements are downloading. This feature lets the user start reading the text while the rest of the information is being received.
- Indicate the percentage of a page that remains to be retrieved. This information is often displayed at the bottom of the screen and helps the user make a decision about whether to wait before trying to access the remainder of the information.

Types of World Wide Web Browsers

Now that you know what a Web browser is and what features are necessary and desirable, let's look at the various types of browsers available. We'll limit our discussion to those most commonly used.

LYNX AND OTHER NONGRAPHICAL BROWSERS

There are two main categories of World Wide Web browsers: nongraphical and graphical. Nongraphical browsers, such as **Lynx** for IBM or IBM PC compatibles from the University of Kansas, only display text and are not suitable for multimedia since the other elements of multimedia (audio, graphics, and full-motion video) cannot be displayed or played. Although text is the only element that can be displayed, it may be possible to download graphics, audio, and video files that can be used in other multimedia players and programs. Figure 2-1 shows a Lynx screen.

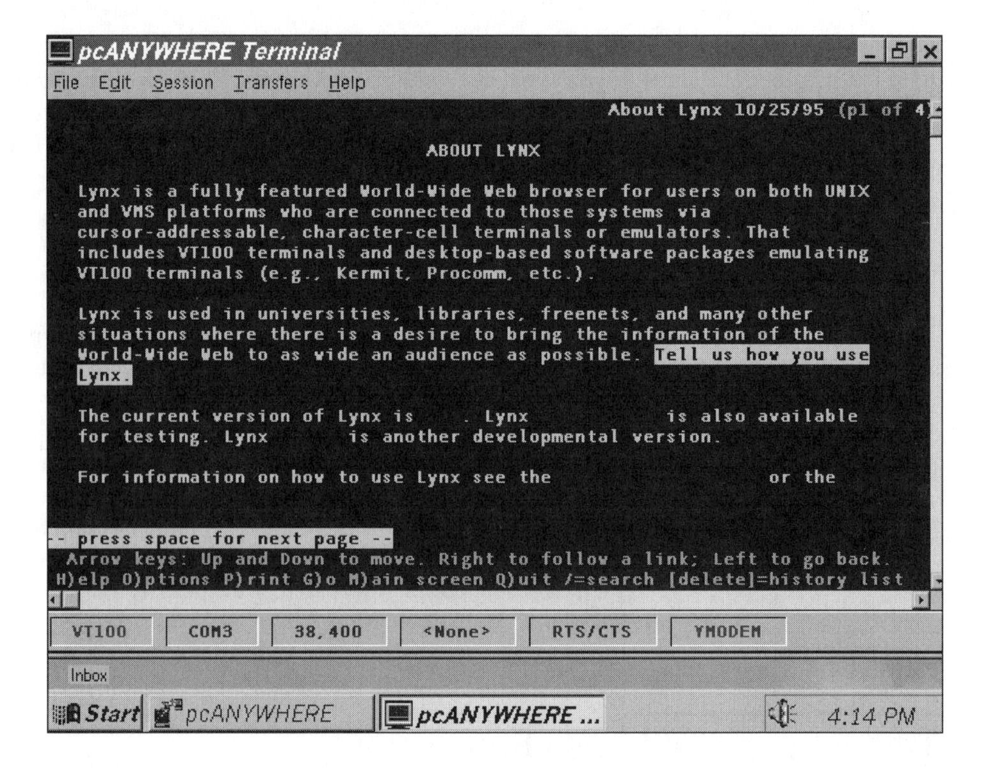

Figure 2-1. The Lynx Browser

GRAPHICAL BROWSERS

The discussion in this textbook will be limited to graphical browsers that have full multimedia capabilities and can display or play text, audio, graphics, and full-motion video. Graphical browsers can be divided into two subcategories: those that are free and those that can be purchased from a vendor. Purchasing a browser gives the user a legal right to use it, and the vendor will usually provide product and technical support. This may be important to a novice user who would not otherwise have anyone to turn to when problems occur or questions arise. Many software vendors provide access to an Internet provider as an option to the purchaser.

Let's examine five graphical browsers: Microsoft Internet Explorer, NCSA Mosaic, Cello, WinWeb, and Netscape. Each of these work well, but some possess more desirable features than others. Which of these have you heard about or have already used?

Microsoft Internet Explorer If you are using Windows 95, you may already have the Microsoft Internet Explorer available on your desktop. This browser is available by purchasing the Microsoft Plus! software, an add-on package for Windows 95. It is also available for free download at http://www.microsoft.com/ie/default.asp. There was much controversy when Windows 95 was first issued because the software's manufacturer, the Microsoft Corporation, provided a means of automatically connecting to the Microsoft Network, an

Internet service provider. Several commercial online services took legal action to try to block Microsoft from offering the Microsoft Network automatically to anyone installing Windows 95. In the end, the courts allowed Microsoft to issue the Web browser as planned. The Microsoft Internet Explorer, free to all users, is based on NCSA Mosaic (see the next section) and distributed under a licensing agreement with Spyglass, Inc. One of the features of this browser's version 3.0 is the toolbar's buttons, which light up when the mouse pointer is over them. Another feature is the ability to open a Favorites folder of frequently visited sites. Figure 2-2 shows the Microsoft Internet Explorer.

NCSA Mosaic Some people incorrectly use the term *Mosaic* to describe all graphical browsers. In fact, NCSA **Mosaic** is a specific browser that was developed by the National Center for Supercomputing at the University of Illinois at Urbana-Champaign as the first graphical browser available for UNIX and Macintosh machines. It is largely because of Mosaic that the World Wide Web gained the popularity that it enjoys today. The browser was developed by students and made available to the public for free, and it is still being upgraded on a regular basis. Although commercial versions of Mosaic may be purchased, a free version continues to be available.

The latest Windows version of Mosaic offers a hotlist option that lists a number of predefined *starting points* that users can use to link to popular Web sites. Starting points are arranged in categories such as Entertainment, Government,

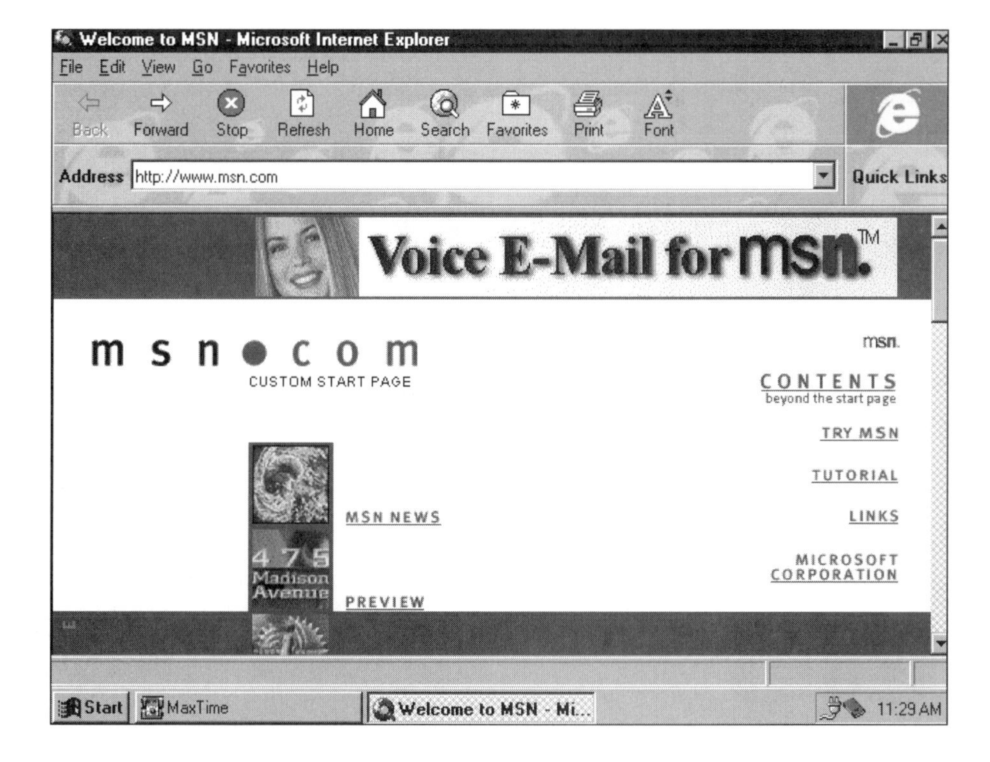

Figure 2-2. The Microsoft Internet Explorer Browser

News, and Search Engines. Another feature is the so-called Audio Events, in which a recorded voice in a sound file announces certain features as they are selected if the user's computer has a sound card and speakers.

The NCSA Mosaic IBM and IBM PC–compatible browser version 2.0 or greater can be downloaded from the ftp.ncsa.uiuc.edu server with the /PC/Windows/Mosaic path or by using the following URL: ftp://ftp.ncsa.uiuc.edu/PC/Windows/Mosaic. Select the subdirectory representing the version of Windows you are operating. If you are using Windows 3.1, you may also need to download and install OLE Enabled WIN32s 1.2 or greater driver at ftp://ftp.ncsa.uiuc.edu/Mosaic/Windows/Win31x/Win32s. Figure 2-3 shows the NCSA Mosaic browser screen. The Macintosh NCSA browser (MacMosaic) can be downloaded by using the following from the host: ftp.ncsa.uiuc.edu with /Mosaic/Mac as the path. See Appendix A for instructions on downloading files. Among the commercial versions of Mosaic is the Spry Mosaic. This browser can be purchased, but it also is available to CompuServe subscribers as a free add-on to the Windows CompuServe Information Manager (WinCim). Subscribers simply download a copy of an installation program called Netlauncher from CompuServe. The WinCim dialer is configured so that new phone numbers and modem setup do not need to be entered. Figure 2-4 shows the Spry Mosaic browser.

Figure 2-3. The NCSA Windows Mosaic Browser

Figure 2-4. The Spry Mosaic Browser

Cello A cello is a musical instrument, but Cello is also a graphical browser developed about the same time as Mosaic. Designed at Cornell University's Legal Information Institute by Thomas R. Bruce, this browser is available only for the IBM or IBM PC–compatible computer. Its features are all accessed via pulldown menus. Unlike Mosaic and the other graphical browsers, Cello does not contain a toolbar. The only clickable buttons are the Stop and Home buttons. However, it is a good alternative for computers with limited disk space. With Cello's default setting, hypertexted text is shown with a dotted rectangle rather than with an underline. It is possible to customize the appearance of Cello's screen. For example, hypertexted text can be displayed as a dotted underline and screen background colors may be changed. Like the other Web browsers, Cello provides a bookmark feature. The bookmarks are alphabetically URL homepage names. The feature lets you copy, delete, and dump the list to a file. Cello may be downloaded from ftp.law.cornell.edu using the /pub/LII/Cello path or by using the following URL: ftp://ftp.law.cornell.edu/pub/LII/Cello. Figure 2-5 shows a Cello screen. See Appendix A for instructions on downloading files.

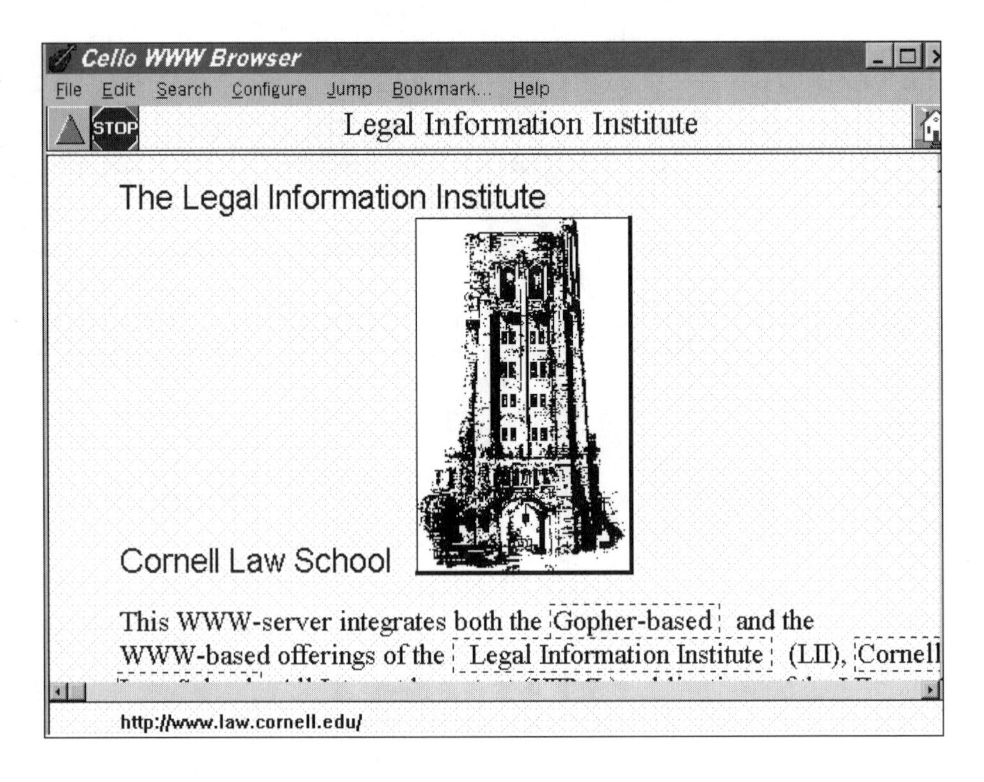

Figure 2-5.
The Cello Browser

WinWeb WinWeb, a free graphical browser from EiNet (TradeWave Corporation), is not as popular as the other graphical browsers. Although it's similar to NCSA Mosaic, WinWeb possesses some differences. One difference is that a dialog box appears during linking to a hyperlink that shows the document is being loaded. (Most other browsers show this information at the bottom of the screen.) One of WinWeb's disadvantages is the fact that the user does not know when transmission is delayed; other browsers show some type of moving graphic, usually at the top of the screen, so that the user knows that the link is still active. The browser screen has a toolbar, pulldown menus, and a hotlist feature.

The WinWeb browser may be downloaded from the info.forthnet.gr server using the /pub/infosystems/www/einet/pc/winweb path or by using the following URL: ftp://info.forthnet.gr/pub/infosystems/www/einet/pc/winweb. Figure 2-6 shows a WinWeb browser screen. A Macintosh version of this browser (MacWeb) can be downloaded from the info.forthnet.gr server with the /pub/infosystems/www/einet/macweb path. See Appendix A for instructions on downloading files.

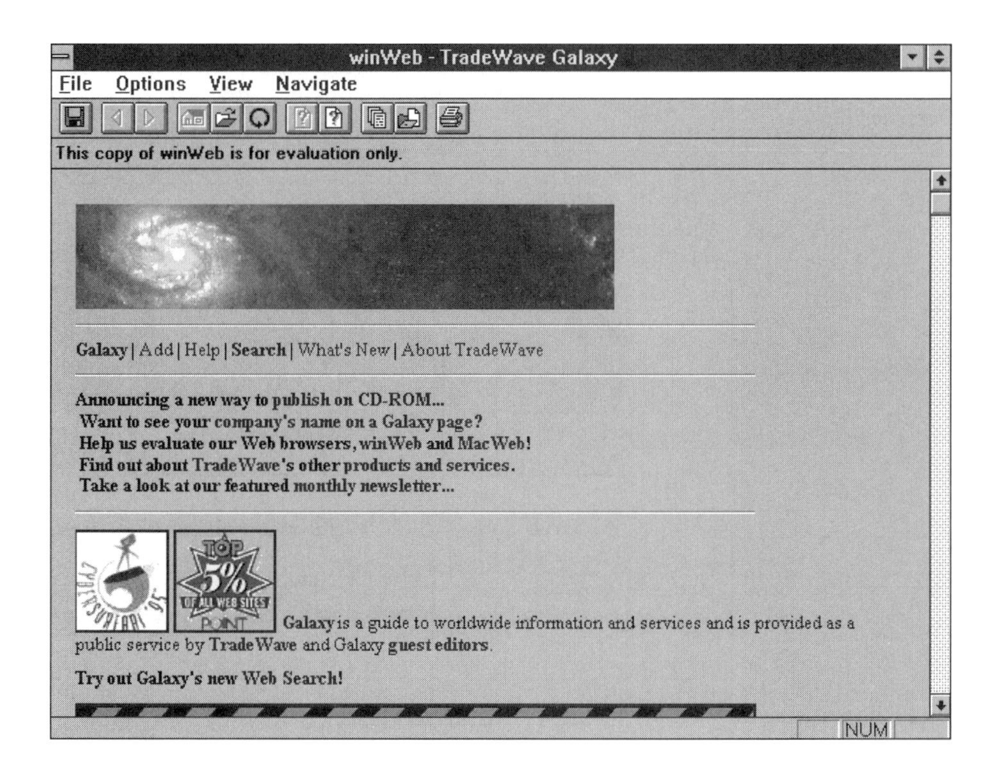

Figure 2-6. The WinWeb Browser

Netscape Navigator If you haven't heard about Netscape, you've either been living in a cave or in an Internet-free zone. Netscape was developed by Marc Andreessen, who worked on a prototype for the NCSA Mosaic browser. He and other fellow University of Illinois students decided to start a new company that would provide Web software. The **Netscape Navigator** (see Figure 2-7) is currently the most frequently used graphical Web browser. A recent research study found that 84 percent of all Web users were using Netscape, and browser statistics indicate that Netscape was used by 74.4 percent of those persons accessing the link.

Because of its wide use, all discussions of multimedia applications in this textbook will assume that you are using Netscape. It is available for both the IBM or IBM PC–compatible and Macintosh platforms, so the description of this browser applies to users of both computer types. Netscape is available free to students and faculty, but others must pay a registration fee. Printed documentation is available for an additional fee.

Figure 2-7.
The Windows Netscape Browser

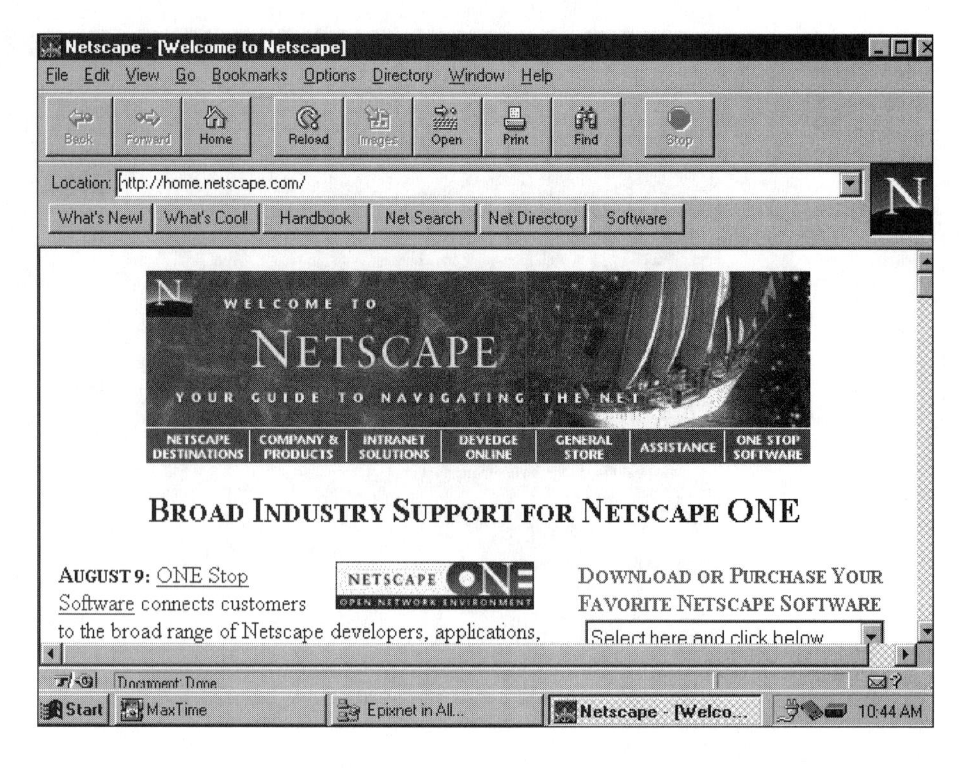

Netscape Features

Why is Netscape so popular? The latest versions contain the following superior features:

- Netscape provides visual clues during the downloading process. The program indicates when it has connected to a site and displays the percentage of bytes that have been downloaded as well as the percentage and number of bytes left to be downloaded. It also indicates when the browser has completed the download.
- Netscape phases in the graphics features of the Web page being downloaded. In Netscape 3.0 a graphic box outline of the image appears first and updates are made until the entire image is downloaded.
- Netscape recognizes special extensions to the HTML Web language that enhance the appearance of the pages after they are downloaded. For example, one extension allows backgrounds of different colors and textures to be viewed.
- Netscape lets users send and receive electronic mail messages as well as view Usenet newsgroups.

- Netscape supports interactive multimedia content such as Java Applets, frames, and inline plug-ins.
- Using state-of-the-art message encryption and digital signatures, Netscape possesses automatic facilities for assigning digital IDs that let users conduct online financial transactions and send and receive e-mail and newsgroup messages securely.
- Netscape includes such features as client-side image mapping and support for multiple simultaneous streaming of video, audio, and other data formats as well as support for the Progressive JPEG file format.
- Netscape's Frame Feature is a sophisticated page-presentation capability that allows the display of multiple, independently scrollable frames on a single screen, each with its own distinct URL.

Steps and Procedures for Using Netscape

Now that you know about Netscape's features, let's discuss how to install and use this browser. Unless Netscape has already been downloaded and installed for you, you must use the following steps and procedures for downloading, installing, and starting it. If the program is already installed in your computer, you can skip ahead to "Starting Netscape."

DOWNLOADING NETSCAPE

If you are using an IBM or an IBM PC compatible, you must first use file transfer protocol (ftp) to download the EXE file labeled either as n16 or n32 from Netscape Communication Corporation's ftp server at ftp.netscape.com with the /pub/navigator/ path. The n16 file is a 16-bit version for use with Windows 3.1. You may also need to download, extract, and install the pw1118.exe file from ftp.microsoft.com with the path /softlib/MSLFILES to be able to run this version. The n32 file is a 32-bit version for use with Windows 95 and later versions. Choose the subdirectories of the version you desire.

Once you have downloaded either of these files, place it in a temporary subdirectory on your hard drive. If you are using a Macintosh, you can also download a copy of Netscape from the ftp.netscape.com server. See Appendix A for instructions on downloading files.

INSTALLING NETSCAPE

To install Netscape, follow these simple steps:

1. With the extracted files in a subdirectory of your hard disk, use the Windows command Run to run the self-executing file. The file loads the setup program and the automated setup procedure will be initiated. A dialog box (see Figure 2-8) will be displayed. Click on Next.

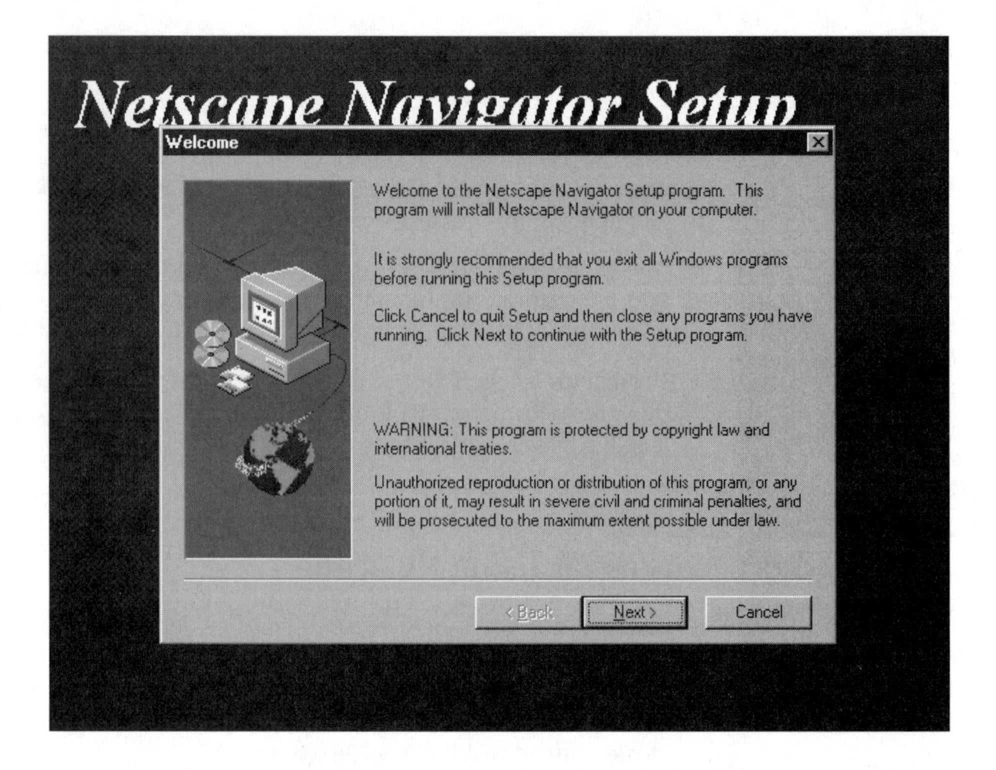

Figure 2-8. The Netscape Setup Dialog Box (Windows 95)

2. When you are prompted for the hard drive location, indicate the name of the subdirectory where you want the program files to reside by clicking on Browse. A default subdirectory will be displayed (i.e., C:\NETSCAPE or C:\PROGRAM FILES\...), and you may accept this one or choose another. The setup program will create the new subdirectory for you. The program files will be copied into the subdirectory that you indicated.

3. You may be prompted to enter the name of the Windows group in which you want to add the Netscape program icon. A default group (Netscape) will be provided, and you may accept this one or choose another, such as Internet Tools.

4. When the installation program has finished copying the files, you should find the files that you need to run the program in the specified subdirectory. You will also find a new program group that you indicated in step 3. If you are using Windows 95, a new folder will be created on your desktop. Check to see that the subdirectory with the files is on your hard disk and that the new Windows group or folder appears when you click on Start/Programs.

STARTING NETSCAPE

To start Netscape, follow these steps:

1. If you are using Windows 3.1 and have a direct Internet connection, you may need to load the TCP/IP drivers, such as Trumpet WinSock. This will not be necessary for Windows 95, which has a built-in TCP/IP driver.
2. If you are using Windows 3.1, open the Netscape group window if it is not already open. Double-click on the Netscape icon to load the program (see Figure 2-9). If you are using Windows 95, click on the Start button in the bottom left corner of the screen and hold down the mouse button. Point the mouse cursor on Programs, then point the cursor on the Netscape group. Finally, point the cursor on the Netscape icon and release the mouse button to start the program (see Figure 2.10).

Figure 2-9. Netscape Icon in Windows 3.1

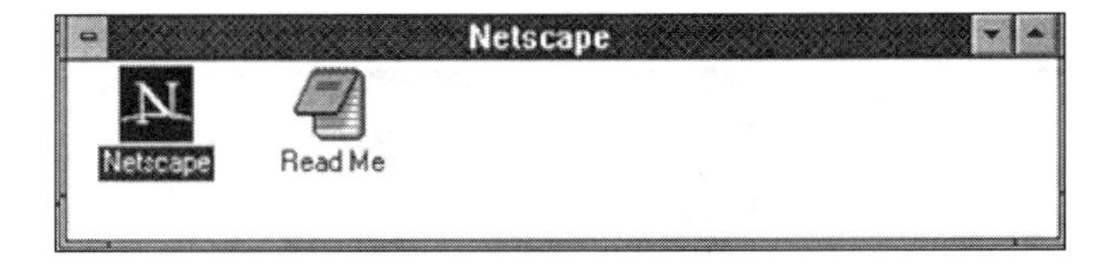

Figure 2-10. Netscape Icon in Windows 95

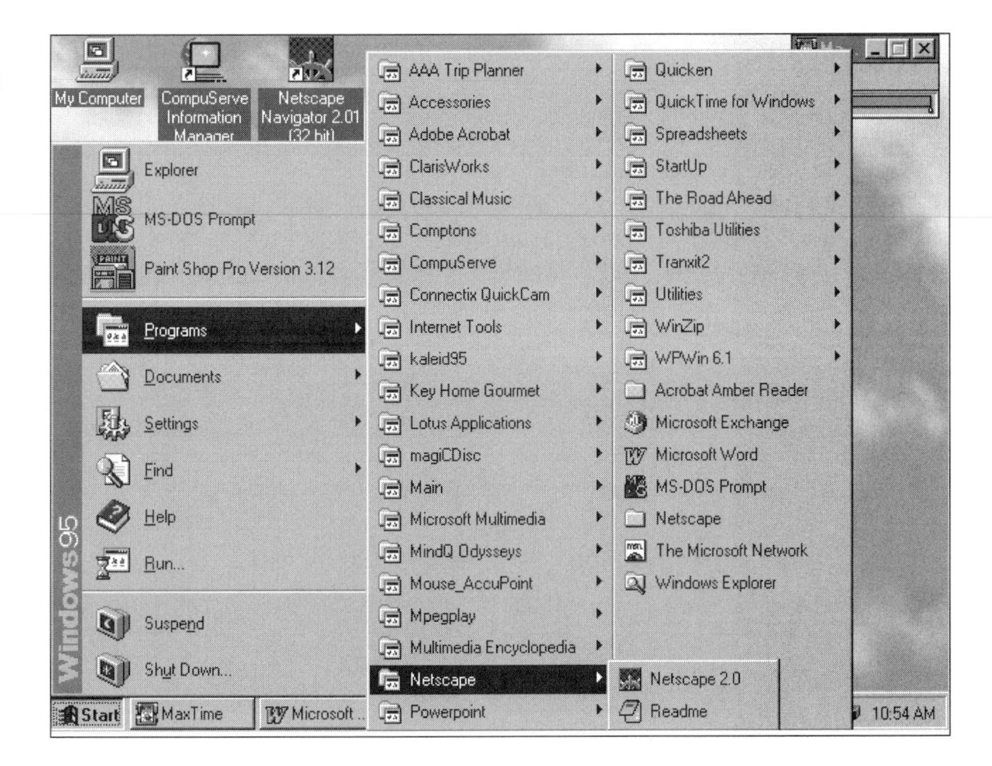

3. The first time you start Netscape, you will be prompted to read the licensing agreement. Click to accept the agreement.

Parts of the Main Netscape Screen

If you followed the steps in the previous section, you now have the user-friendly Netscape browser on the screen before you. Let's continue by identifying the parts of the main Netscape screen. Refer to Figure 2-11 to identify each part as it is explained.

- *Menu bar.* The menu bar can be used to access most Netscape features.
- *Toolbar.* The toolbar, located directly below the menu bar, contains commonly used commands that can be turned off and on. The toolbar can be turned off if you desire.
- *Location box.* The location box, located directly below the toolbar, displays the current URL you are browsing. You can also type a URL in this area by clicking an insertion point in the box. You can delete a displayed URL by selecting the URL and pressing the Delete key or by clicking an insertion point to the right of the URL and backspacing.

Figure 2-11. Parts of the Main Netscape 3.0 Screen

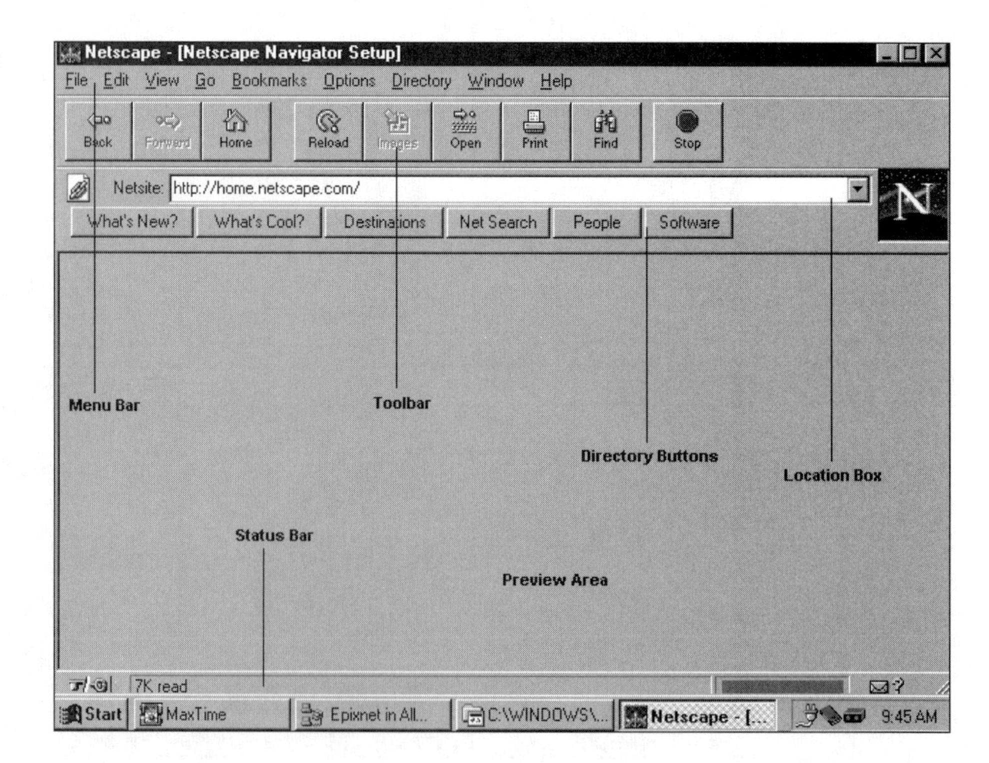

- *Preview area.* The preview area is the portion of the main screen in which you view Web pages after they have been downloaded.
- *Status bar.* The status bar at the bottom of the screen shows information about the progress of a download. It will display the URL for a targeted hyperlink and the number of bytes and percentages that have been downloaded. At the right side of the status bar, a graphic bar shows how much of a document still must be loaded.

Let's now examine several of these parts of the Netscape screen in more detail to gain a better understanding of them.

MENU BAR

The menu bar is a means of accessing the main features of Netscape 2.0 or 3.0. If you're using an earlier or later version of Netscape, some of the menu bar items will be different. Individual menu items can be pulled down by moving the mouse cursor on the menu title and then clicking the left mouse button or by pressing the Alt key and the underlined letter of the menu title. Not all menu items are available at all times. Items that are not available will appear in gray screen rather than in black. An item with an ellipsis (...) will open a dialog box after you click on it. Individual menu items can be selected by clicking the mouse cursor on them or by pressing the Alt key and the underlined letter of the menu item. A menu item can also be accessed by using the shortcut keys that appear to the right of some items.

- *File.* The File menu contains items that allow you to open a WWW location by entering a URL, open a stored disk file that contains a URL, save a downloaded Web page, compose and mail an electronic mail message, change the page setup defaults, print a downloaded Web page, preview a downloaded Web page, close the program, and exit the program.
- *Edit.* The Edit menu contains items for undoing a previous command; cutting, copying, or pasting text to the Windows Clipboard, and finding text that appears in a downloaded Web page in the preview area of the screen.
- *View.* The View menu contains items for reloading a document or a cell, loading images, refreshing the link, indicating the document source that provides a view of the HTML programming language that supports the page, and document information. When selected, the document information appears in a split screen: the top of the screen shows the structure of the document and the bottom provides information about the document.
- *Go.* The Go menu contains items for moving backward and forward to access previously linked documents. Selecting Home in this menu will take you back to the homepage. Selecting Stop Loading will halt the loading of the current link. A list of previous links appears at the bottom

of the pulldown menu. Clicking on each of these will take you back to the page representing the link.

- *Bookmarks.* The Bookmarks menu allows you to add documents to a file after you link to them. Selecting Go to Bookmarks in this menu takes you to a screen on which all of your previously saved bookmarks are displayed. You are able to edit and view the makeup of the saved bookmarks through pulldown menus.
- *Options.* The Options menu provides a means for changing the settings and page appearance of Netscape. Choosing General Preferences, Mail & News Preferences, Network Preferences, and Security Preferences will take you to specific setup pages where you can change default settings. There are also a series of items on this menu for settings pertaining to the location of the toolbar, directory positions, auto loading images, and showing FTP file and document encoding. Clicking on each item will turn the setting on as indicated by a checkmark. The Document Encoding item allows you to change settings to receive documents in various languages including Japanese, Chinese, and Korean. The Save option will save all options in a file.
- *Directory.* The Directory menu contains items for accessing helpful and interesting locations on the Web that Netscape maintains. Options include Netscape Home, What's New, What's Cool, Customer Showcase, and Internet Search.
- *Window.* The Window menu contains options that allow different screens to be displayed. These include a new Netscape window and the Netscape Mail, News Reader, and Bookmark screens. The History option will display a list of previously linked Web pages. A list of the previous links appears at the bottom of the pulldown menu.
- *Help.* The Help menu provides access to online help provided by Netscape. The About option at the top of the menu is the only item that does not require you to be online to use.

TOOLBAR

The toolbar consists of nine buttons: Back, Forward, Home, Reload, Images, Open, Print, Find, and Stop. These buttons represent another way to access commonly used commands on the menus. Because the toolbar is a convenient way to use the functions, it's a good idea to display it at all times. Now let's look at each of the buttons on the toolbar.

The Back and the Forward buttons allow you to move through links you have made during the current session. When the Back button is clicked, Netscape will take you to the preceding link.

Clicking on the Forward button moves you to the next link. Clicking each of these buttons more than once will take you back or forward one additional link with each click.

The Home button takes you back to the page you have identified as your home or starting page during the current session.

The Reload button will reaccess and reload the currently linked page. This is useful when for some reason the page does not load correctly.

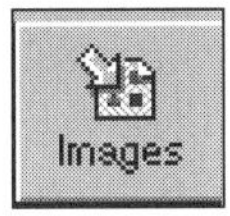

The Image button indicates whether or not the Auto Load Images menu item on the Options menu has been selected or deselected. You may want to deselect this option to load pages with graphic images faster. This is especially beneficial when you are using a dial-up Internet connection or when a page loads slowly because of heavy Internet traffic.

The Open button can be used for entering a new URL that Netscape will access. An alternate method is to choose Open Location on the File menu. When you click on the Open button, the Open Location dialog box appears (see Figure 2-12).

Click on OK after keying the URL in the dialog box.

Figure 2-12. The Open Location Dialog Box

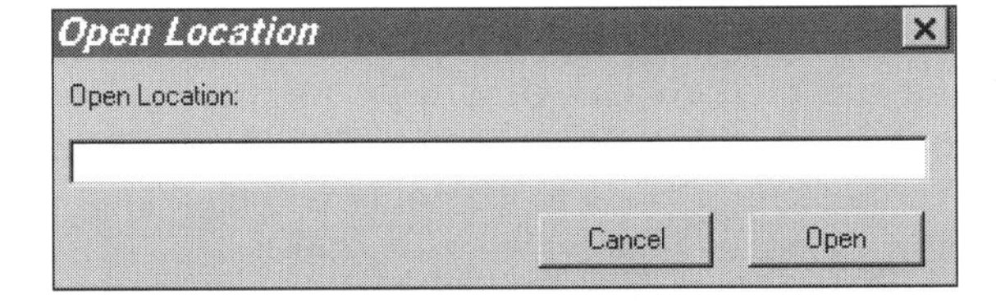

Clicking on the Print button sends the current displayed page to a connected printer. An alternate method is to choose Print on the File menu. When you click on the Print button, the Print dialog box appears. (See Figure 2-13 for a Windows 95 Print dialog box.) Indicate the name of the printer, print range, number of copies, and click on OK.

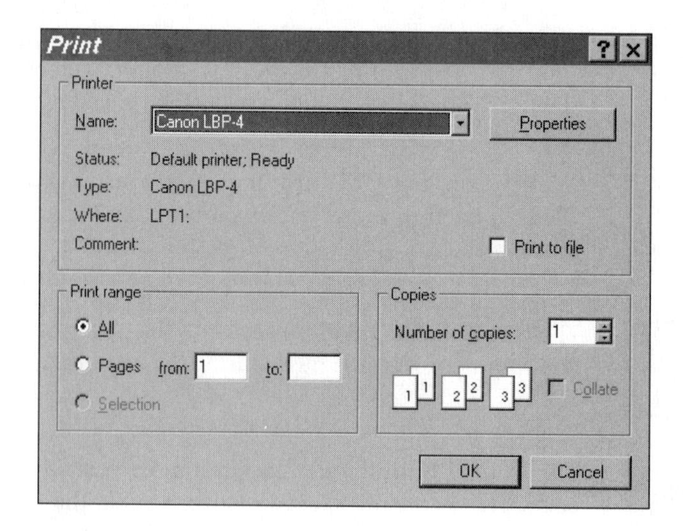

Figure 2-13. The Print Dialog Box for Windows 95

The Find button is used to search the currently displayed page for a word or phrase. After clicking on the Find button, the Find dialog box appears (see Figure 2-14). By clicking on the Up or Down radio buttons (circles) in the dialog box, you indicate in what direction you want the search to go. Click on Find Next to execute the search. Keep clicking to find additional words or phrases. Click on Cancel to close the dialog box.

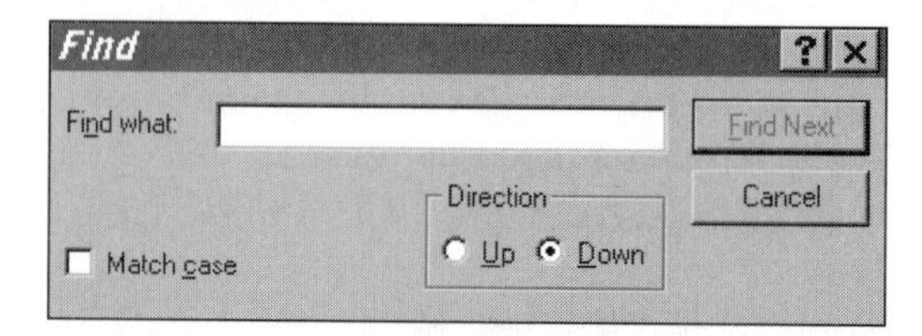

Figure 2-14. The Find Dialog Box

Clicking on the Stop button will abort the current attempt to download a page. You may want to use this button when you decide that you no longer want to continue downloading a page because of slow response time or other reasons.

DIRECTORY BUTTONS

The row of buttons located under the Toolbar are the Directory buttons. Each of these buttons will link you to pages maintained by Netscape. The following information is provided by clicking on each of these buttons. Newer versions of Netscape will contain new and different buttons.

What's New? — The What's New! button will display a page that provides information about Netscape Communications.

What's Cool! — The What's Cool! button will display pages about interesting places to link to on the WWW.

Destinations — The Destination button will display links that help you to find information on the Web.

Net Search — The Net Search button will display pages that link you to various search engines to find Web pages on the Internet. A more detailed description of searching on the WWW is provided later in this chapter.

People — The People button will display links to search engines for finding electronic mail (e-mail) addresses.

Software — The Software button will take you to a page for configuring Netscape to make it more helpful and powerful. This includes a means for registering Netscape and an easy way to select new components, plug-ins, and helper applications.

Setting Netscape's Preferences

Defaults are the presettings a program displays after it is installed. In Netscape 2.0 or 3.0 you can set preferences on the Options menu in the categories General, Mail and News, Network, and Security.

- *General.* The General option permits you to change the general appearance of Netscape, how it links to Web pages, and how it utilizes multimedia features of the Web.
- *Mail and News.* The Mail and News option allows you to set options for receiving and sending mail and for using the built-in newsgroup reader.
- *Network.* The Network option provides information about available cache space, network connections, and proxies.
- *Security.* The Security option allows you to indicate how you want to be warned about insecure transactions.

Let's examine each of these setting categories more closely so that you can gain a better understanding of them.

THE GENERAL OPTION

The General option has the following groups of settings related to multimedia: Appearance, Fonts, Colors, Images, and Helper Applications. The Appearance group allows you to change the appearance of the toolbar, indicate the home-page location you want to link to first, and indicate the link styles (see Figure 2-15).

The Helper Applications group is especially pertinent to multimedia because it allows you to configure external programs to view or play multimedia file formats (i.e., sound, graphics, video, etc.). Although some of the file types are supported by built-in players and viewers, others need to be installed and set up by the user. This group allows you to prescribe which player and viewer you want Netscape to use when it encounters various file types on the Web. Many are available for free download. Figure 2-16 shows a Helper Application group page in which the RealAudio player has been installed for file types with the extensions ra and ram. See Chapters 4 and 9 for the specific steps needed to install a helper application.

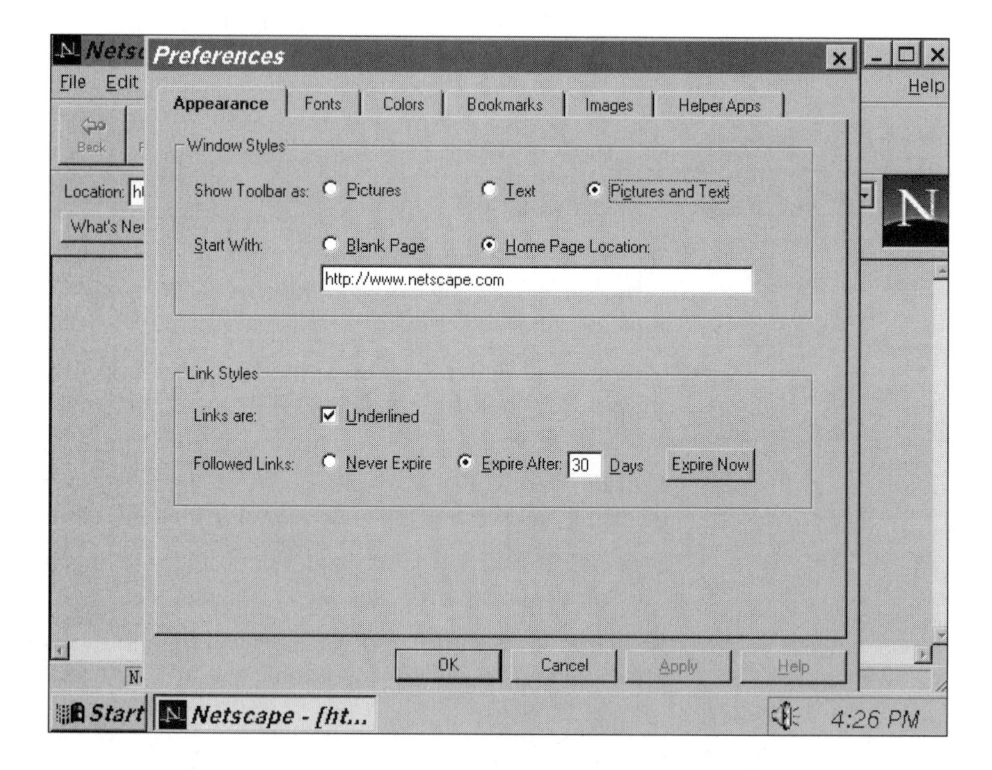

Figure 2-15. The Appearance Preferences Settings

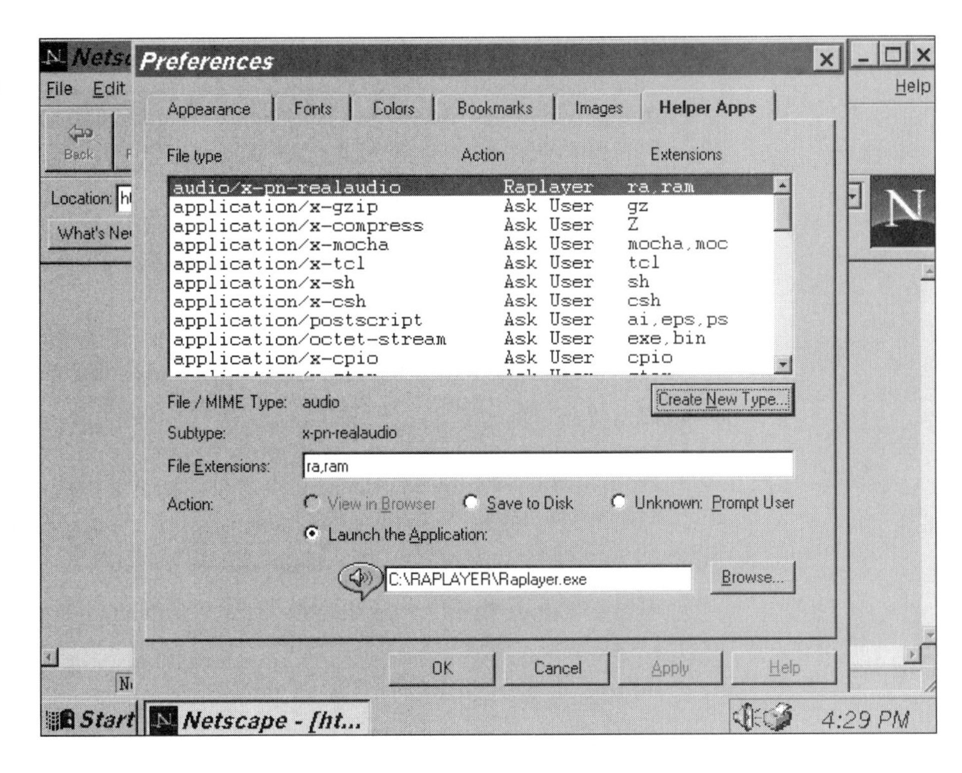

Figure 2-16. Helper Application Group

Searching on the World Wide Web

How to find things on the Web is the first question anyone asks about Netscape or any other Web browser. The three mechanisms on the Web for researching a specific topic are search engines (or spiders), metasearch engines, and resource (or subject) trees. Each will produce similar but slightly different results.

SEARCH ENGINES (SPIDERS)

Search engines (also known as **spiders**) allow you to search by title or headers of documents, by words or phrases in the documents themselves, or by other indexes or directories. On Netscape these search engines can be accessed by pressing the Net Search button. The following search engines were available when this text was being written. You can access each engine by either clicking on the hyperlinked name using the Net Search button or by clicking the Open button on the toolbar and entering the URL indicated.

- *Excite* (http://www.excite.com). The Excite database contains more than 1 million Web documents as well as the preceding two weeks of Usenet newsgroup messages and classified advertisements.

- *Yahoo* (http://www.yahoo.com). Yahoo's search engine can search the Yahoo site, Usenet, and e-mail addresses. You can configure the search to find new listings added over time and to display various numbers of listings per page.
- *InfoSeek Net Search* (http://www2.infoseek.com). InfoSeek Net Search is a fast, accurate, and comprehensive way to search the Web. This engine allows you to enter a question or as many words or phrases that you need to describe what you want to find.
- *Lycos* (http://www.lycos.com). The Lycos engine searches for Web documents by document title, headings, links, and keywords.
- *Magellan* (http://magellan.mckinley.com). Magellan provides a powerful search engine that includes Web sites, ftp and gopher servers, newsgroups, and telnet sessions. The sites are viewed and rated by a group of professionals and academics.

To use Excite to conduct a search, follow these steps:

1. Access Excite by clicking the Net Search button and the Excite tab.
2. Click an insertion point in the white search bar next to the Search button.
3. Key in the word or words you want to use in your search.
4. Change the options below the search bar, if desired.
5. Click on the Search button. (Your browser should connect to the Excite site and the results of your search should be returned.)
6. Link to any of the hypertexted titles by clicking on it.

The following search engines are also available:

- *Configurable Unified Search Interface (CUSI)* (http://cuiwww.unige.ch). The CUSI engine searches a large number of Web engines for documents, people, software, dictionaries, and other items.
- *World Wide Web Worm* (http://wwww.cs.colorado.edu/wwww). This engine searches for Web pages by matching specific hypertext, page titles, or words with the Web pages.
- *Internet Sleuth* (http://www.isleuth.com). This is a comprehensive database of Web databases. A search request generates a list of hyperlinks to other search engines. You can also browse through Sleuth's database via an alphabetically arranged list.

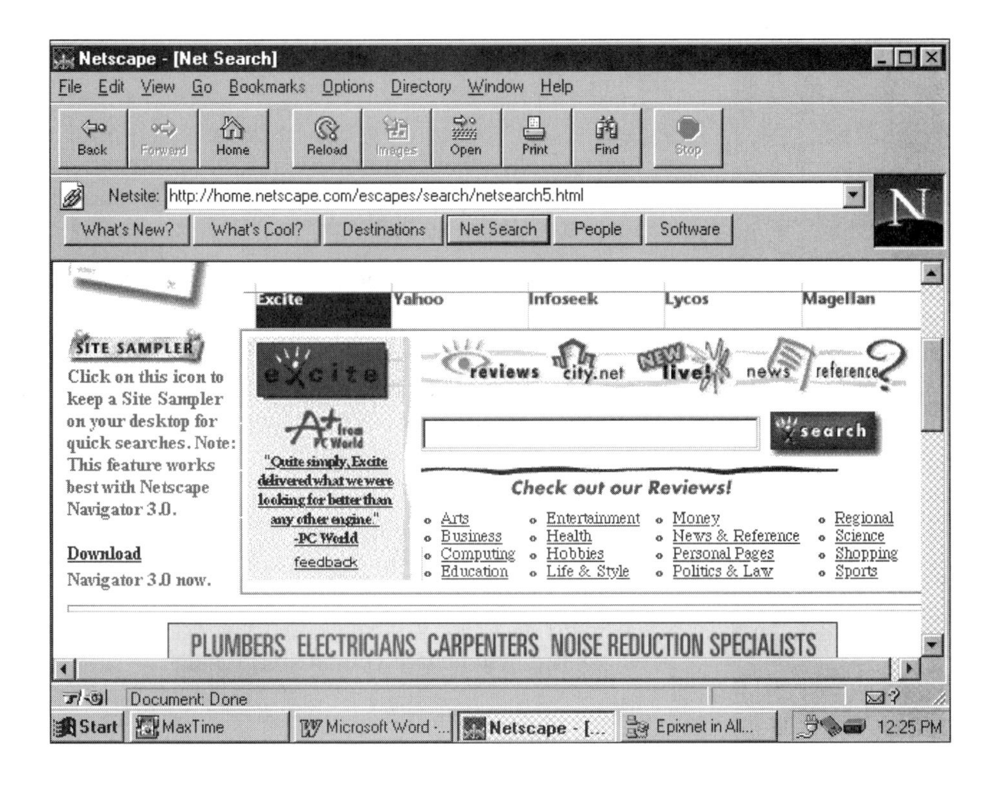

Figure 2-17. Excite

METASEARCH ENGINES

Metasearch engines search multiple search engines simultaneously for the entered term. MetaCrawler, for example, is a multithreaded search site that can take your search terms and simultaneously feed them to nine sites: Open Text, Lycos, WebCrawler, InfoSeek, Excite, Inktomi, Alta Vista, Yahoo, and Galaxy. It organizes the results into a uniform format and displays them. MetaCrawler also includes the option of sorting the hits into lists in a number of different ways, including by locality, region, and organization. MetaCrawler can even eliminate invalid URLs. The URL is http://metacrawler.cs.washington.edu. Additional metasearch engines may be found at http://www.internic.net.

WEB SEARCH STRATEGIES

The following strategies for searching the Web are suggested by Bryan Pfaffenberger in his book *Web Search Strategies* (MIS Press, 1996).

1. *Understand the search engine's limitations.* Does it include gopher, ftp, and archie information or just Web (http) documents? Understanding what's in the database will help you devise more effective search terms.

2. *Access the search engine.* Don't try to search during peak usage hours (12 p.m. to 3 p.m. Eastern Standard Time).
3. *Type one or more search words in the text box.* Start with fairly specific terms. Type the most important words first. Choose search options, if any are available. Search different parts of Web documents: titles, document content, hyperlinked text.
4. *Click the Start Search or Submit button.* Clicking this button initiates the search.
5. *View the list of retrieved documents.* Most search engines rank the search results numerically, giving the first document the highest score. If you find a document that looks like it's relevant, click the cited hyperlink.
6. *Refine and repeat the search, if necessary.* Common problems include too many or too few documents.

Pfaffenberg offers these additional tips:

Check your spelling. Make sure that you don't have any incorrect spellings that will negate your search results.

Don't use common articles or Web terms. Articles or terms to avoid are and, the, and http.

Don't type plurals. Always type the singular form of all search words for hyperlinked main topics.

RESOURCE (OR SUBJECT) TREES

An alternative way of finding resources on the Web is to use a **resource** or **subject tree**. Web sites are listed by topic and subtopic and also may have a search utility that helps you to search for topics within the listed categories.

- *Yahoo* (http://www.yahoo.com) **Yahoo**, which we already used as a search engine, stands for "Yet Another Hierarchically Odoriferous Oracle" and is one of the best of the subject trees available on the Web. The main topics in Yahoo at the time this text was written were Arts, Business and Economy, Computers and Internet, Education, Entertainment, Government, Health, News, Recreation, Reference, Regional, Science, Social Science, and Society and Culture. Developed by two students at Stanford University, the service is updated every five days and also features a search utility. To use Yahoo, either enter a word or words into the search bar and click on the Search button or click on one of the hyperlinked main topics. Figure 2-18 shows Yahoo.

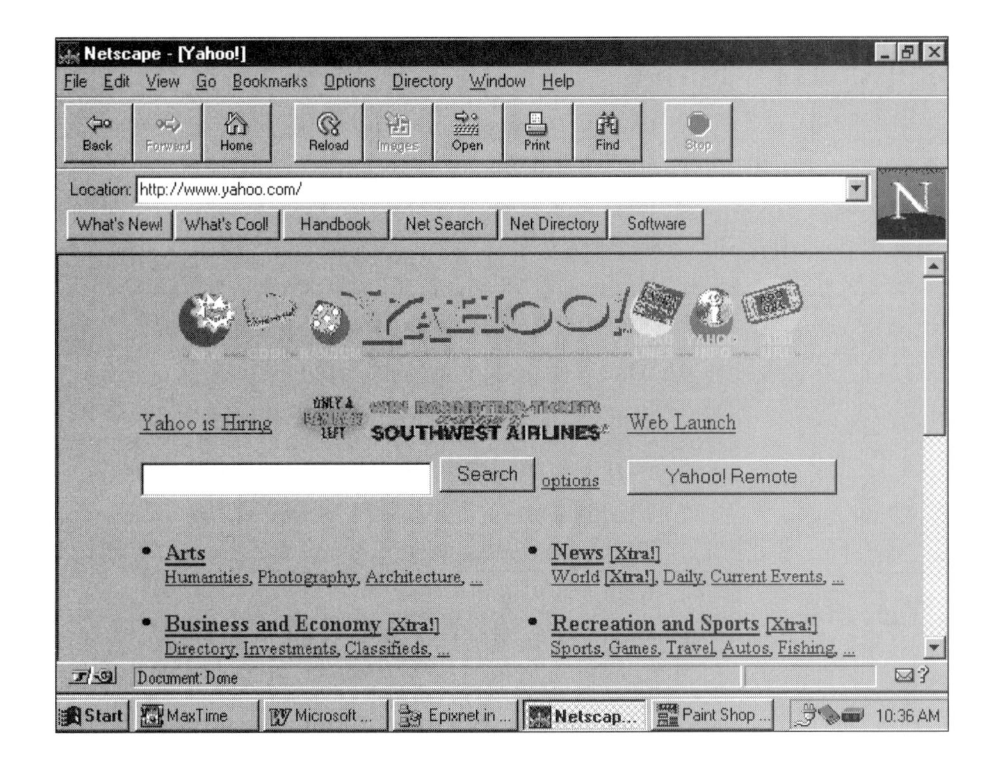

Figure 2-18. Yahoo

- *Scott Yanoff's Internet Services List.* (http://www.spectracom.com/islist). A 1995 award winner of the top 5 percent of Web sites, this service covers the following main topics: Agriculture, Art, Astronomy, Aviation, Biology, Botany, Business/Economics/Financial, Chemistry, College Prep., Computers, Consumer/Commercial Information/Resources, Education/Teaching/Learning, Employment, Food/Recipes/Cooking, FTP, Games/Fun/Chat, Geophysical/Geographical/Geological, Gopher, Government/Politics, History, Internet, Law, Literature/Books/Languages, Math, Medical/Health, Movies, Museums/Laboratories/Exhibits/National Parks, Music/Sounds, News/Electronic Journals/Magazines, Paranormal/Occult/Spiritual/Astrology, Physics, Religion, Software, Space, Sports and Recreation, Television, Theater/Drama, Travel, User Lookup Services/Whois Services, Weather/Atmospheric/Oceanic, Women/Feminism, and WWW/HTML/Mosaic. This list may have changed slightly since the writing of this textbook.

Questions for Review

1. What is the World Wide Web? Why is it used for multimedia applications?
2. What significant events occurred as the World Wide Web was being developed?
3. What are the advantages of the World Wide Web compared with the other Internet services (i.e. ftp, gopher, WAIS)?
4. What is hypermedia? What is hypertext?
5. What is the hypertext markup language (HTML)?
6. Who uses the World Wide Web?
7. What is a uniform resource locator (URL)? What types of URLs are there?
8. What is a Web browser?
9. What are the three basic features of Web browsers? What desirable features should they have?
10. What are the main categories of Web browsers? What is the main difference between them?
11. What are the main graphical Web browsers? How are they different? Which browser is currently the most frequently used?
12. What are the superior features of the Netscape browser?
13. What are the steps necessary to download Netscape?
14. What are the steps necessary to start Netscape?
15. What are the functions of each of the following parts of the main Netscape screen: menu bar, toolbar, location box, preview area, and status bar?
16. How do you set the preferences in Netscape?
17. How do you conduct a search on the World Wide Web? What is a search engine?
18. What are the steps necessary to find resources using Excite?
19. What is a subject tree?
20. What are the steps necessary to find resources using Yahoo?

Exercises

Using the Netscape browser, complete the following exercises.

1. Use the Open Location feature and key in and link to http://home.netscape.com/home/whats-new.html or click on the What's New directory button. Describe what's new at Netscape Communications.
2. Use the Open Location feature and key in and link to ftp://sumex-aim.stanford.edu/info-mac/comm/info/ftp-primer.txt. Save and print the file. How does this document define anonymous FTP?
3. Use the Open Location feature and key in and link to gopher://marvel.loc.gov. How many menu items appear?
4. Using the Excite search engine (http://www.excite.com), conduct a search with the keyword Multimedia. Choose one or more of the hypertexted lists and link to them. What additional information did you find about multimedia?
5. Using the Yahoo subject tree (http://www.yahoo.com), link to Multimedia under Computers and Internet. Next, link to Hypermedia. How many listings are there? Link to one of these items. What information did the site provide?

CHAPTER

3

Hypertext Markup Language (HTML)

As we have seen, hypertext markup language (HTML) is what makes it possible for the Web to support multimedia. This chapter examines how the Web browsers use HTML to process information. You will learn the basic commands to create Web pages with multimedia elements using popular HTML editors.

What You Will Learn

- How HTML is defined
- Advantages and disadvantages of HTML
- How to use HTML
- Types of HTML software utility programs
- Steps and procedures for using Microsoft Internet Assistant
- Steps and procedures for using WordPerfect Internet Publisher
- Steps and procedures for using HTML Writer
- Steps and procedures for using HTML Pro

HTML Defined

The basis of pages that appear on the World Wide Web is the hypertext markup language (HTML), which transmits the structure of documents between users of the Web. Web browsers such as Netscape read HTML files stored in standard ASCII on Web servers connected to the Internet. Unlike word processors, which require programmers to concentrate on formatting and how the document will appear when it is printed, HTML requires programmers to think in terms of rules that will result in a specific document content. HTML has been standardized to prevent confusion among users. Currently, version 2.0 is being used, but a prototype of version 3.0 has been developed and Netscape 2.0 and 3.0 are able to read it. Netscape is also able to read in specially developed HTML extensions that give pages a special appearance.

Advantages of HTML

Although it possesses some limitations, *hypertext markup language* has a few advantages, including these:

1. It is easy to learn; HTML codes do not have to be memorized.
2. HTML is compatible with any Web browser.
3. HTML files are small and take up little storage space.
4. Using HTML, it is possible to create and define a numbered list displayed with bullets or Roman or Arabic numbers.

The first advantage—that you don't need to memorize HTML codes—is possible because a number of editors can be used to create the HTML files containing the codes. This chapter will discuss the types of editors available and will explain how to use several of them. The second advantage—HTML's compatibility with any Web browser—means that anyone with an IBM PC–compatible computer or a Macintosh computer, an Internet connection, and programs such as Netscape or Mosaic can read a Web page created with HTML. The third advantage—that HTML files are small and take up very little storage space compared with other types of program files—means that HTML files do not need to be compressed to fit on a floppy disk. The fourth advantage—that an HTML author can use the language to create and define a numbered list displayed with bullets or Roman or Arabic numerals—illustrates the structured nature of HTML, which provides users with a great deal of flexibility in viewing a document.

Disadvantages of HTML

Nothing is perfect, and HTML is no exception. Some of its limitations include the following:

1. HTML lacks "WYSIWYG" capabilities.
2. It is limited to predetermined codes.

The first disadvantage of HTML, the missing WYSIWYG ("what you see is what you get"), means that while you are composing a document, you can't see exactly how the document will appear when it is viewed with a Web browser. The only way to see the document as it will actually appear is to open the file with a Web browser. This is often a time-consuming and frustrating process for someone trying to develop a Web document. Some HTML editors, however, have a built-in test or preview feature that allows the editor to connect to a Web browser you have configured in advance.

Because HTML limits you to the predetermined codes currently available, you are unable to create any special layout or design effects for which no codes have been written. HTML version 3.0 contains many more new codes than previous versions, however, and subsequent versions will probably contain many more codes, resulting in Web pages with varied layouts.

Using HTML

HTML uses **tags**, instructions to the Web browser about how to display text and other elements. Tags are constructed by enclosing the identifier in less than (<) and more than (>) symbols (basic format: <identifier>) that separate them from the rest of the text. Because the tags are not case sensitive, they can be in lower case, upper case, or a mix of lower or upper case. Thus, all of the following tags would be acceptable: <html>, <HTML>, or <HtMl>. Some HTML editors use upper-case tags, others use lower-case. This textbook will use upper-case tags to make them stand out from text.

There are two basic types of tags: single-element and symmetric. *Single-element tags* consist of one tag only, such as <P>. *Symmetric tags* are used in pairs. One of these can be seen as an On code and the other as an Off code. A tag without a slash (i.e., <HTML>) is the On code, while the tag with the slash (i.e., </HTML>), is the Off code. Required information, including text and other tags, is placed between the On and Off codes.

Hypertext markup language is composed of six specific types of tags. Each of the commands are of equal importance and have a specific purpose. The HTML file created is a text file with either an .htm extension with a MS-DOS computer or an .html extension with a UNIX-based computer. With a Macintosh computer, which does not use file extensions, the HTML file is saved without an extension.

The six types of tags used with HTML are structural, paragraph-formatting, character-formatting, list-specification, hyperlink, and multimedia tags. Each type of tag has a specific purpose and use. Depending on the type of document you are preparing with HTML, you may not need to know all of the possible variations of tags in each type. Some tags in each category are single-element tags, whereas others are symmetric.

Here is a sample HTML document file. We will refer to it as we discuss the various types of HTML tags.

```
<HTML><HEAD><TITLE>Sample HTML Document File</TITLE></HEAD>
<BODY><H1><I>"Using Multimedia Tools and Applications on the
Internet"</I></H1>
<BR><B><CENTER>by Dennis O. Gehris, Ed.D.</B><A
HREF="http://www.bloomu.edu/departments/beoa/welcome2.au">
<BR>Click here for an audio welcome.</A></CENTER>
<HR>
<H2>The text contains the following chapters:</H2>
<BR>
<OL>
<LI>Introduction to the Internet and Multimedia
<LI>The World Wide Web (WWW)
<LI>The Hypertext Markup Language (HTML)
<LI>Graphic Images on the Internet
<LI>Audio on the Internet
<LI>Video and Animation on the Internet
<LI>Computer Conferencing on the Internet
<LI>Videoconferencing on the Internet
<LI>Adobe Acrobat, ASAP WebShow, and Macromedia Shockwave on the Internet
<LI>Virtual Reality on the Internet
</OL>
<HR>
<IMG SRC="http://www.bloomu.edu/departments/beoa/lsmiley.gif">For more
information, contact
<ADDRESS>Integrated Media Group, Wadsworth Publishing
Company, Belmont, CA 94002</ADDRESS> or
visit the group's Web site at
<A HREF="http://www.thomson.com/rcenters/img/img.html">
http://www.thomson.com/rcenters/img/img.html</A>.
</BODY></HTML>
```

STRUCTURAL TAGS

Structural tags identify a file as a HTML document and provide information about the data in the HTML file. The following are structural tags: <HTML> </HTML>, <HEAD></HEAD>,<TITLE></TITLE>, and <BODY></BODY>. The sample

HTML document shows correct order for these tags: <HTML> <HEAD> <TITLE>Sample HTML Document File</TITLE></HEAD><BODY></BODY> </HTML>.

<HTML></HTML>	This tag informs the Web browser what kind of document it is so that it can be displayed properly. The On code is placed at the beginning of the document and the Off code at the end of the document.
<HEAD></HEAD>	This tag allows the Web browser to discover information about the document.
<TITLE></TITLE>	The information that appears between these On and Off codes will appear on the title bar of the window and is used for index information by Web search engines.
<BODY></BODY>	The information that appears between these On and Off codes represents the main part of your document. A special body tag extension that Netscape will read allows you to specify the background color for your document. The extension is placed in the On body code. The format is <BODY BACKGROUND="http://www.domain.name/path/file-name.gif">. Netscape maintains a sampler of body background files at http://home.netscape.com/assist/net_sites/bg/backgrounds.html.

PARAGRAPH-FORMATTING TAGS

Paragraph-formatting tags specify paragraphs and heading levels. The paragraph-formatting tags are <P></P>, <PRE></PRE>, <BR>, <HR>, <H1></H1>, <H2></H2>, <H3></H3>, <H4></H4>, <H5></H5>, and <H6></H6>.

<P></P>	The paragraph tag is used to designate the beginning of a paragraph. HTML version 2 does not require an Off code, but version 3 does. Without this tag, Web browsers would not be able to tell when a paragraph begins and ends, because pressing the Enter key in an HTML editor has no effect on the line endings.
 	The line break tag allows you to break a line without adding a space between the lines. It does not require an Off code.
<HR>	The horizontal rule tag instructs the Web browser to create a horizontal line across the width of the display window.
<H1...6></H1...6>	Each of the six header tags represent different type styles and sizes. The type style of the heading will vary depending on the browser used. Figure 3-1 shows the appearance

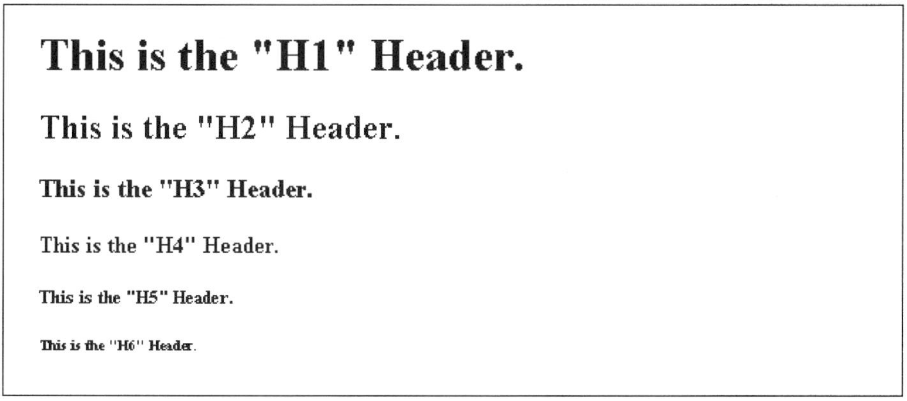

Figure 3-1. Text Using the Six HTML Header Tags with Netscape

This is the "H1" Header.

This is the "H2" Header.

This is the "H3" Header.

This is the "H4" Header.

This is the "H5" Header.

This is the "H6" Header.

of text in each of the six header styles with Netscape. The sample HTML document file uses an H2 code, as follows:

```
<H2>The text contains the following chapters:</H2>.
```

<PRE></PRE>

The *preformatted paragraph tag* is used to create a block of text in a particular format. If you have a table with columns of text or numbers, you can preserve it by using this tag, which is also useful for displaying charts, graphs, and diagrams. The text will appear in the Courier font in most browsers. Here is an example of the <PRE></PRE> tag.

```
<BODY>
<H1>Sales Report</H1>
<H3>Here is the sales report for September, October, and November.</H3>
<PRE>
Sales Report
                 Sept.          Oct.           Nov.
                 ______         ______         ______
Jones            394.23         223.83         256.55
Smith            212.45         323.21         423.33
Yuber            563.33         343.22         443.56

</PRE></BODY></HTML>
```

Figure 3-2 shows how the file appears in the Netscape browser.

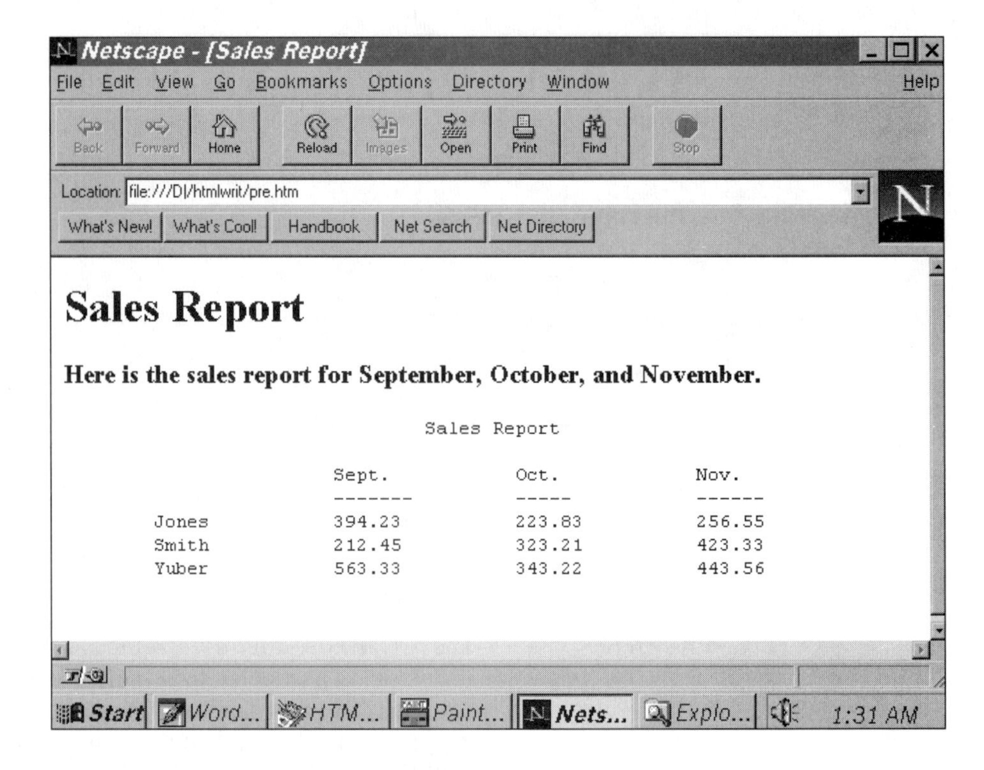

Figure 3-2. Preformatted Text Displayed in Netscape

CHARACTER-FORMATTING TAGS

Character-formatting tags allow for the application of various type styles (such as bold, underline, or italics) to the characters in documents. The most popular character-formatting tags are <B></B>, <U></U>, <I></I>, <CENTER> </CENTER>,<TT></TT>, <EM></EM>, <STRONG></STRONG>, <CITE></CITE>, and <ADDRESS></ADDRESS>. There are also special character commands. For example, because HTML uses the < and > characters to start and end a tag, < is used for a less than (<) character and > for a greater than (>) character. Some of the other special character commands are & for an ampersand (&), and " for double quotes ("). There are several obscure character-formatting tags, including: <SAMP></SAMP>, <CODE></CODE>, <KBD></KBD>, <VAR></VAR>, and <DFN></DFN>. The discussion here will focus on the most commonly used tags. Figure 3-3 shows text that uses each of these tags in Netscape.

<B></B>	The *bold tag* permits you to boldface text.
<U></U>	The *underline tag* underscores text.
<I></I>	The *italics tag* produces italicized text.
<CENTER></CENTER>	The *center tag* horizontally centers text placed between the On and Off codes. It can also be used to center a hyperlinked element. The sample HTML document file uses this tag as follows:

Figure 3-3. Text Using the Most Popular HTML Character-Formatting Tags with Netscape

This text illustrates the "B" tag.
This text illustrates the "I" tag.
This text illustrates the "U" tag.
This text illustrates combining tags: **bold**, *italics*, underline.
`This text illustrates the "TT" tag.`
This text illustrates the "EM" tag.
This illustrates the "STRONG" tag.
This text illustrates the "CITE" tag.
This text illustrates the "ADDRESS" tag.

```
<CENTER>by Dennis O. Gehris, Ed.D.</B><A
HREF="http://www.bloomu.edu/departments/beoa/welcome2.
au"> <BR>Click here for an audio welcome.</A></CENTER>
```

<TT></TT> — The *typewriter text tag* will display text in the Courier font in most Web browsers.

<EM></EM> — Like the italics tag, the *emphasis tag* will produce italicized text.

<STRONG></STRONG> — Like the bold tag, the *strong tag* will produce boldfaced text.

<CITE></CITE> — The *citation tag* can be used for a citation for a title or a reference. The text will appear in italics.

<ADDRESS> </ADDRESS> — The *address tag* is used for an electronic mail address and/or to identify the author of a document or the Webmaster. The text will appear in italics. The sample HTML document file uses this tag as follows:

```
<ADDRESS>Integrated Media Group, Wadsworth Publishing
Company, Belmont, CA 94002</ADDRESS>
```

LIST-SPECIFICATION TAGS

List-specification tags allow you to format several different types of lists: ordered (numbered), unordered (bulleted), directory (for short lists), menu (with each item one line long), and glossary (definition). Glossary lists are those in which each item has a term and a definition. Creating a list involves starting the section with the correct On command as shown below. In every type of list except a glossary, the <LI> tag is used before each item. After all items are keyed in the list, the Off command is used. The sample HTML document file uses the following list-specification tags:

```
<OL>
<LI>Introduction to the Internet and Multimedia
<LI>The World Wide Web (WWW)
<LI>The Hypertext Markup Language (HTML)
<LI>Graphic Images on the Internet
<LI>Audio on the Internet
<LI>Video and Animation on the Internet
<LI>Computer Conferencing on the Internet
<LI>Videoconferencing on the Internet
<LI>Adobe Acrobat, ASAP WebShow, and Macromedia Shockwave on the Internet
<LI>Virtual Reality on the Internet
</OL>
```

<UL></UL> This is the tag for an *unordered list,* a list with bulleted items that can be arranged in any order. Other tags, such as those for bold and hypertext, can be included within the list. The following is an example of the HTML tags for an unordered list. Figure 3-4 shows how this file would be displayed in Netscape.

```
<HTML><HEAD><TITLE>Example of an Unordered List</TITLE></HEAD>
<BODY>
<P><H1>Example of an Unordered List in HTML</H1>
<P><H3>The Types of HTML Lists</H3>
<UL>
<LI>Unordered Lists
<LI>Ordered Lists
<LI>Directory Lists
<LI>Menu Lists
<LI>Glossary Lists
</UL></BODY></HTML>
```

Figure 3-4. Unordered List in Netscape

Example of an Unordered List in HTML
The Types of HTML Lists
• Unordered Lists
• Ordered Lists
• Directory Lists
• Menu Lists
• Glossary Lists

<OL></OL> This is the tag for an *ordered list,* in which the items are preceded with numbers. The items also do not need to be in any particular order and will be numbered consecutively. You shouldn't number the items because numbers will be displayed automatically when you use a browser. The following is an example of the HTML tags for an ordered list; Figure 3-5 shows how this file would be displayed in Netscape.

```
<HTML><HEAD><TITLE>Example of an Ordered List</TITLE></HEAD>
<BODY>
<P><H1>Example of an Ordered List in HTML</H1>
<P><H3>My Shopping List</H3>
<OL>
<LI>Bread
<LI>Milk
<LI>Eggs
<LI>Cheese
<LI>Margarine
</OL>
</BODY></HTML>
```

Figure 3-5. Ordered List in Netscape

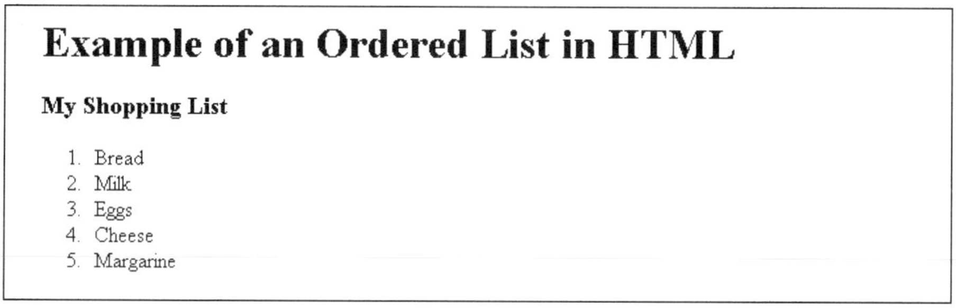

Example of an Ordered List in HTML

My Shopping List

1. Bread
2. Milk
3. Eggs
4. Cheese
5. Margarine

<DIR></DIR> *Directory lists* and *menu lists,* variations of unordered lists, are intended for lists with short items that can be displayed in a compact style.

<MENU></MENU> *Menu lists* usually contain no bullets, numbers, or other labels. However, some Web browsers display menu and directory lists with bullets.

<DL></DL> *Glossary (definition) lists* are intended for lists of terms and their corresponding definitions. Each list item contains a term and its definition. The term is preceded with the <DT> tag and the definition is preceded with the <DD> tag. The following is an example of a glossary list; Figure 3-6 shows how this file would be displayed in Netscape.

```
<HTML><HEAD><TITLE>Example of a Glossary (Definition) List</TITLE></HEAD>
<BODY>
<P><H3>Example of a Glossary (Definition) List in HTML</H3>
<P><H4>Internet/Multimedia Definitions</H4>
<DL>
<DT><B>Internet</B>
<DD>A network of computer systems that are interconnected in about 130
countries.
<DT><B>Multimedia</B>
<DD>The integration of at least two media: text, audio, graphics, and full-motion
video.
<DT><B>World Wide Web</B>
<DD>A wide-area hypermedia information retrieval initiative aiming to give
universal access to a large universe of documents.
</DL>
</BODY></HTML>
```

Figure 3-6. Glossary List in Netscape

Example of a Glossary (Definition) List in HTML

Internet/Multimedia Definitions

Internet
A network of computer systems that are interconnected in about 130 countries.
Multimedia
The integration of at least two media: text, audio, graphics, and full-motion video.
World Wide Web
A wide-area hypermedia information retrieval initiative aiming to give universal access to a large universe of documents.

HYPERLINK TAGS

Hyperlink tags, which lay down the specifications related to moving from one hyperlink to another on the Web, make it possible to link from a place in the present document to another place. You can hyperlink to complete documents or to graphic images, audio clips, and video clips. To hyperlink the spot in your document, you must know the URL of the place or file you want to link to and the text you want to use to create the link. The sample HTML document uses the following hyperlink:

```
or visit the group's Web site at
<A HREF="http://www.thomson.com/rcenters/img/img.html">
http://www.thomson.com/rcenters/img/img.html</A>.
```

When you use the correct hyperlink tags, the hyperlinked text appears in a different color and is usually underlined in most browsers. There are three kinds of hyperlinks:

- *Local hyperlink:* those that jump to a place inside your file.
- *Remote hyperlink 1:* those that link to another file on another computer.
- *Remote hyperlink 2:* those that jump to a named location in another file.

Each kind of hyperlink uses the <A></A> anchor tag, but in a slightly different manner.

Local Hyperlink

```
<A HREF="#TargetName">Linked Word(s)</A>
```

This is the format of the hyperlink tag that will jump to a specific place in the same document. #TargetName would be replaced with the actual name of the target, which is marked with the next tag. The Linked Words would be replaced with the actual hyperlinked text.

```
<A NAME=#TargetName></A>Target Location
```

This is the hyperlink tag for the location to which to jump. #TargetName would be replaced with the actual name of the target, which is marked with the next tag. The Linked Word(s) would be replaced with the actual hyperlinked text.

Here is an example of a file that uses these two hyperlink tags:

```
<HTML><HEAD><TITLE>Example of Local Hyperlink</TITLE></HEAD>
<BODY>
<P><H3>Example of a Local Hyperlink in HTML</H3>
<A HREF="#address">Click here to jump to the address.</A>
<H3>ABC University offers a wide variety of course offerings in all of the popular
disciplines.</H3>
<A NAME="address"></A>
<ADDRESS>ABC University<BR>
320 University Boulevard<BR>
State University, NJ 08648</ADDRESS>
</BODY></HTML>
```

Remote Hyperlink 1

```
<A HREF="Document's URL">Linked Word(s)></A>
```

This is the format for the hyperlink code that is used to jump to another document. Document's URL would be replaced with an actual URL. Linked Word(s) would be replaced with actual hyperlinked text.

Here is an example of a file that uses this hyperlink tag.

```
<HTML><HEAD><TITLE>Example of Remote Hyperlink #1</TITLE></HEAD>
<BODY>
<P><H3>Example of a Remote Hyperlink #1 in HTML</H3>
<A HREF="http:://www.microsoft.com">Click here to access Microsoft's Web
Page.</A>
```

Continued on following page

Continued from previous page

```
<H3>ABC University offers a wide variety of course offerings in all of the popular
disciplines.</H3>
<ADDRESS>ABC University<BR>
320 University Boulevard<BR>
State University, NJ 08648</ADDRESS>
</BODY></HTML>
```

Remote Hyperlink 2

```
<A HREF="http://www.domain.name/path/filename.html#TargetName">Linked Word(s)
</A>
```

This is the format for the hyperlink code that is used to jump to a remote computer's URL and the name of the target document file. The http://www.domain.name/path/filename.html#TargetName would be replaced with an actual URL, path, filename, and target name. The Linked Word(s) would be replaced with actual hyperlinked text.

Here is an example of a file that uses this hyperlink tag.

```
<HTML><HEAD><TITLE>Example of Remote Hyperlink #2</TITLE></HEAD>
<BODY>
<P><H3>Example of a Remote Hyperlink #2 in HTML</H3>
<A HREF="http://www.abc.edu/faculty/abc.html#faculty">Click to view a listing of
our faculty</A>
<H3>ABC University offers a wide variety of course offerings in all of the popular
disciplines.</H3>
<ADDRESS>ABC University<BR>
320 University Boulevard<BR>
State University, NJ 08648</ADDRESS>
</BODY></HTML>
```

MULTIMEDIA TAGS

Multimedia (or *asset integration*) *tags* provide access to multimedia applications relating to graphical images, video, and audio.

Graphic Image Tags Detailed coverage of graphic images is included in Chapter 4. For now, the syntax for the main HTML tags needed to access graphic images will be sufficient.

```
<IMG SRC="http://www.domain.name/path/sample.gif">
```

This is the tag format for placing a graphic image located on a remote computer on your Web page. If the graphic image file is located on the local server, the http://www.domain.name part of the format can be eliminated. When graphic images are combined with text, you can add an ALIGN= code to the tag so that the graphic is aligned at the top, bottom, or middle of the existing text. For example, if you wanted to indicate a middle alignment, the tag would be as follows:

```
<IMG ALIGN=MIDDLE SRC="http://www.domain.name/path/sample.gif">
<IMG SRC="http://www.domain.name/path/sample.gif"[Textual Description]>
```

This tag format would be used to provide a textual description of the image, which is displayed in text-only browsers.

```
<A HREF="http://www.domain.name/path/file"><IMG
SRC="http://www.domain.name/path/sample.gif"></A>
```

This tag format would be used to designate a graphic image as a hyperlink. By clicking on the graphic, the user jumps to the appropriate hyperlink target.

The sample HTML document file uses the following graphic image tag:

```
<IMG SRC="http://www.bloomu.edu/departments/beoa/lsmiley.gif">.
```

Video and Sound Tags Detailed coverage on video is included in Chapter 6 and on audio in Chapter 5. For now, the syntax for the main HTML tags needed to access video and audio files will be sufficient. The sample HTML document file contains the following sound tag:

```
<A HREF="http://www.bloomu.edu/departments/beoa/welcome1.au">.
<A HREF="file://video.avi">Click here for video</A>
```

This is the tag format for downloading the file video.avi from the local Web server. The Click here for video or other appropriate text would be hyperlinked text, so that when the user clicks on the text the video file would be accessed and an external application would be launched.

```
<A HREF="http://www.server/path/video.avi">Click here for video</A>
```

This is the tag format for downloading the file video.avi from the remote Web server.

```
<A HREF="file://audio.wav">Click here for audio</A>
```

This is the tag format for downloading the file audio.wav from the local Web server. The Click here for audio or other appropriate text would be hyperlinked, so that when the user clicks on the text, the audio file would be accessed and an external application would be launched.

```
<A HREF="http://www.server/path/audio.au">Click here for audio.</A>
```

This is the tag format for downloading the file audio.au from a remote Web server. Table 3-1 lists all of the HTML tags.

Next we'll examine the various types of software utility programs that can be used to prepare HTML documents.

TABLE 3-1 THE HTML TAG TYPES

Types of Tags	Tag Formats
Structural	<HTML></HTML>, <HEAD></HEAD>, <TITLE></TITLE>, <BODY></BODY>
Paragraph formatting	<P></P>, <PRE></PRE>, , <HR>, <H1></H1>, <H2>/</H2>, <H3></H3>, <H4></H4>, H5></H5>, <H6>,</H6>
Character-formatting	<B></B>, <U><U/>, <I></I>, <CENTER></CENTER>, <TT></TT>, <EM></EM>, <STRONG></STRONG>, <CITE></CITE>, <ADDRESS></ADDRESS>
List-specification	<LI>, <UL></UL>, <OL></OL>, <DIR></DIR>, <MENU></MENU>, <DL></DL>, <DT>,<DD>
Hyperlink	<A HREF="#TargetName">Linked Word(s)</A>
	<A Name=#TargetName></A>Target Location
	<A HREF="Document's URL>LinkedWord(s)></A>
	<A HREF="http://www.domainname/path/fileame html#TargetName">Linked Words(s)</A>
Multimedia	<IMG SRC="http://www.domainname/path/sample.gif">
	<IMG SRC="http://www. domainname/path/sample.gif"[Textual Description]>
	<A HREF="http://www.domainname/path/file"><IMG SCR="http://www.domainname/path/sample.gif"</A>
	<A HREF="http://www.domainname/path/video.avi">Click here for video</A.
	<A HREF="http://www.domainname/path/audio.au"<Click here for audio.</A>

HTML Software Utility Programs

Why are HTML software utility programs necessary? Netscape allows you to view the HTML tags in a document that has been loaded by choosing By Document Source from the View menu. However, it is usually not possible to edit the codes from inside the browser. You can use a basic word processing program or a text editor to create text files containing the HTML tags, but this is a time-consuming and tedious procedure. A much better approach is to use some type of software utility program specially designed to produce HTML files. Although these programs are a great help to users producing HTML files, knowledge of HTML is usually required. There are numerous HTML utility programs available for both IBM or IBM PC–compatible and Macintosh computers. Some programs are available for free or require a small shareware fee. Others can be purchased from software companies.

Utility programs fall into two categories: **converters** and **editors**. *Converters* can be used if you have an existing document such as a brochure or newsletter already saved on a file. Converter programs are either standalone or add-ons to an existing word processing or desktop publishing program. Standalone converters convert a particular file format (such as Rich Text Format or Postscript) to HTML. Presently, add-on program converters or templates exist for Microsoft Word for Windows and WordPerfect for Windows. These add-on programs can be acquired for free or for a small fee from the software companies.

New versions of word processing and desktop publishing programs, such as PageMaker, include or will include built-in HTML conversion utilities. HTML editors are used when you are creating a Web document from scratch and if a document does not already exist in any type of file. The discussion here will be limited to the most popular editors available for free or at a small fee for both the IBM or IBM PC–compatible and Macintosh computers.

IBM AND IBM PC–COMPATIBLE ADD-ON EDITORS

Add-on editors let you utilize an existing program you may be currently using as an HTML editor. Let's look at some of the add-on editors available for IBM and IBM–compatible computers.

Microsoft's Internet Assistant and WordPerfect's Internet Publisher Using an add-on editor is the easiest way to produce HTML files because you don't need any knowledge of HTML tags. However, it may still be a good idea to learn the tags in case no add-on editor is available. The Microsoft Internet Assistant editor is available for Microsoft Word 6.0 for Windows and Microsoft Word 7.0 for Windows (Windows 95). Internet Assistant is available for free download at http://www.microsoft.com/msword or via Microsoft's Network. Microsoft Word '97 can be used as a HTML editor and also contains a Web page wizard template that walks you through the steps in creating a Web page.

Users of either of these programs are able to click on a button or access an item from a menu to add HTML elements to a file, which is then converted to HTML format. The Internet Publisher is available for Corel's WordPerfect for Windows 6.1 and 7.0 (Window 95). Internet Publisher comes preinstalled with WordPerfect 7.0. Internet Publisher for WordPerfect 6.1 is available for free download at ftp://ftp.corel.com/pub/WordPerfect/wpwin/61/wpipzip.exe. Your choice of using Internet Assistant or Internet Publisher will depend on the word processing program to which you have access: Microsoft Word or WordPerfect. Obviously, you don't need to utilize both software add-ons.

Netscape Navigator Gold Netscape Navigator Gold is actually the Netscape browser with an add-on editor. You are able to see the document that you are developing exactly as it will appear in Netscape. As with the other add-on editors, Netscape Gold does not require you to learn HTML editing codes; settings, tags, and formatting options let you insert elements such as headers and links. When you are editing a remote HTML file, the HTML file and all of the associated graphic images files will first be saved to a disk on your local computer. A One-Button Publish feature makes it possible to download your completed HTML file to a server.

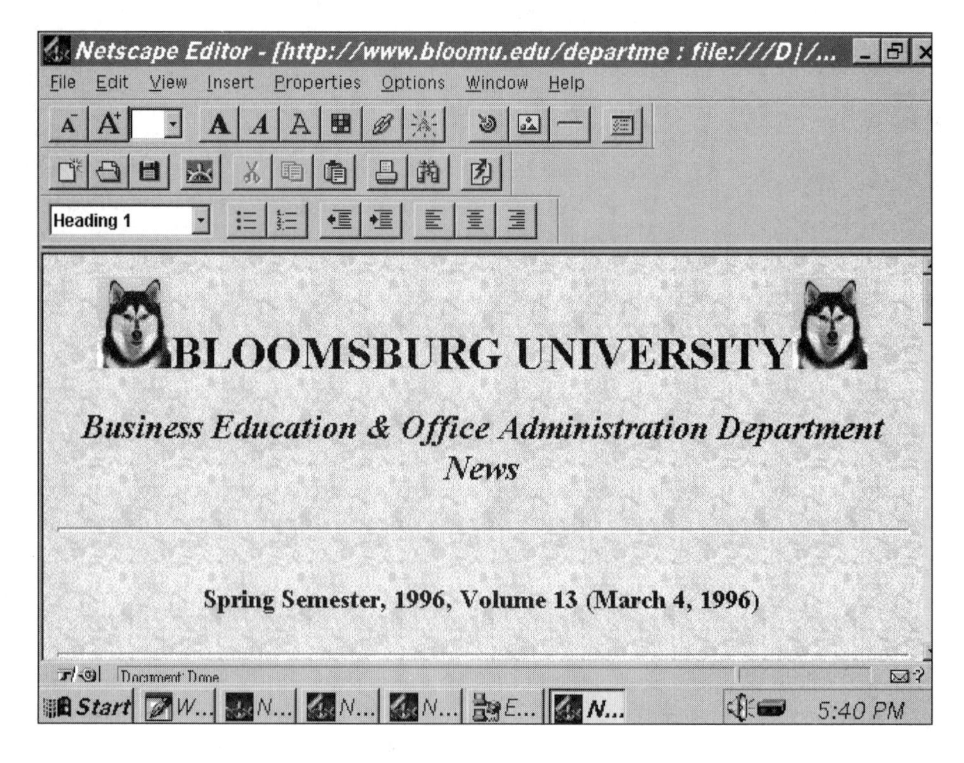

Figure 3-7.
The Netscape Gold HTML Editor

Netscape Page Wizard is Netscape's custom Web page creation tool for quick, high-impact results. Netscape Navigator Gold also contains a Page Starter function offering thirteen templates that can be used to begin developing a homepage by linking to http://home.netscape.com/home/gold3.0_templates.html. Currently, templates exist in the following categories: Personal/Family, Company/Small Business, Department, Product/Service, Special Interest Group, and Interesting and Fun. Figure 3-7 shows the Netscape Gold HTML Editor.

Steps and Procedures for Using Microsoft Internet Assistant

To use Internet Assistant, you must have a copy of Microsoft Word for Windows Version 6.0 or 7.0 installed on your computer. After downloading Internet Assistant, you must install the add-on program. Run the self-executing file and follow the onscreen instructions. One of the features of Microsoft Internet Assistant is a built-in Web browser.

To use Internet Assistant to create a simple HTML file, follow these steps:

1. Run Microsoft Word.
2. Click on Browse Web on the File menu. The Web browser screen should appear (see Figure 3-8).

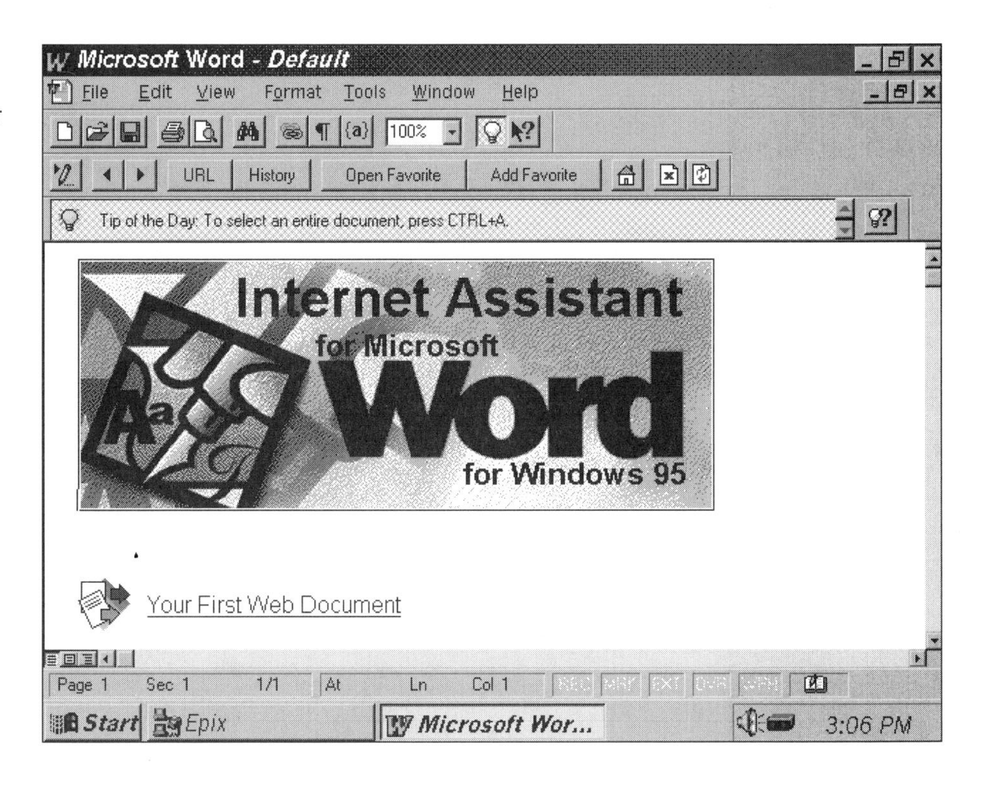

Figure 3-8. The Microsoft Assistant Browser

3. Click on New on the File menu.
4. Click on the HTML icon on the General Tab in the New dialog box. Click on OK. Notice that the formatting toolbar has changed to include HTML buttons (see Figure 3-9).
5. Click on HTML Document Info... on the File menu. Key SAMPLE MICROSOFT INTERNET ASSISTANT DOCUMENT in the HTML Document Head Information dialog box (see Figure 3-10).
6. Key SAMPLE MICROSOFT INTERNET ASSISTANT DOCUMENT as the title. Select the text by dragging the I-beam cursor over the text while holding the left mouse button. Click on the large A on the toolbar to increase the size of the font. Click on the white area to the right of the title to deselect the text. Press the Enter key (see Figure 3-11).
7. Click on Horizontal Rule on the Insert menu. Click on Save on the File menu, keying a:samplehtml as the filename.
8. Key the following text. Do not press the Enter key at the end of the lines.

   This is a sample document created with the Microsoft Internet Assistant add-on to Microsoft Word. Internet Assistant can be downloaded free by linking to the following URL:

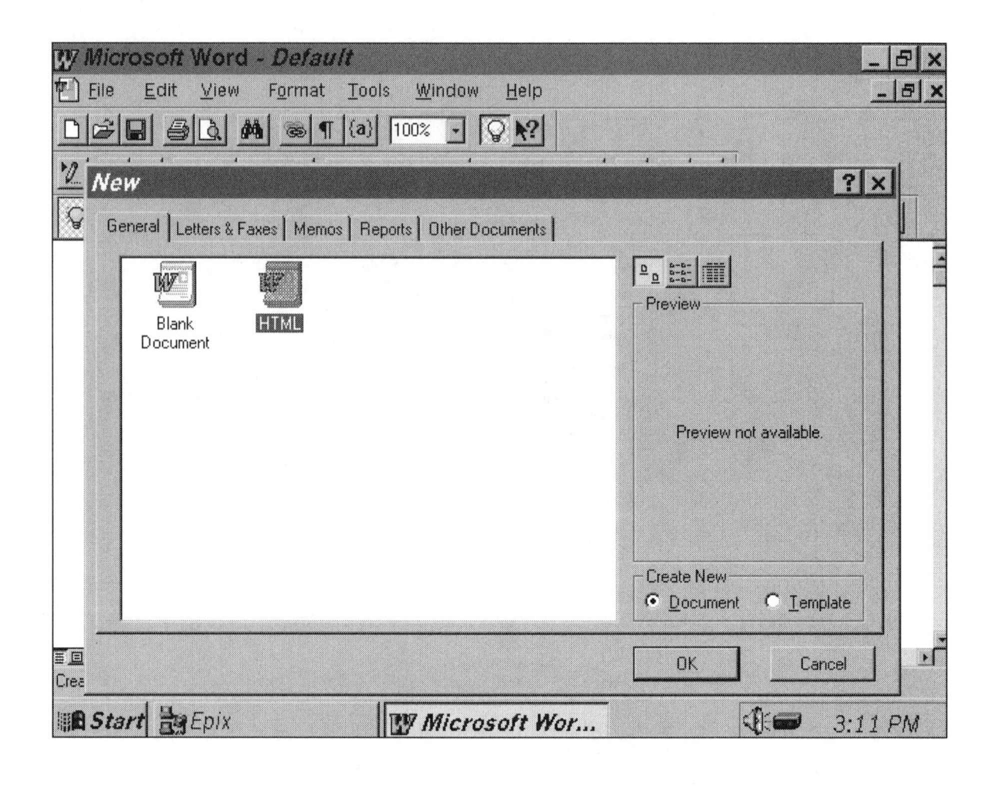

Figure 3-9.
The New Dialog Box in Microsoft Word

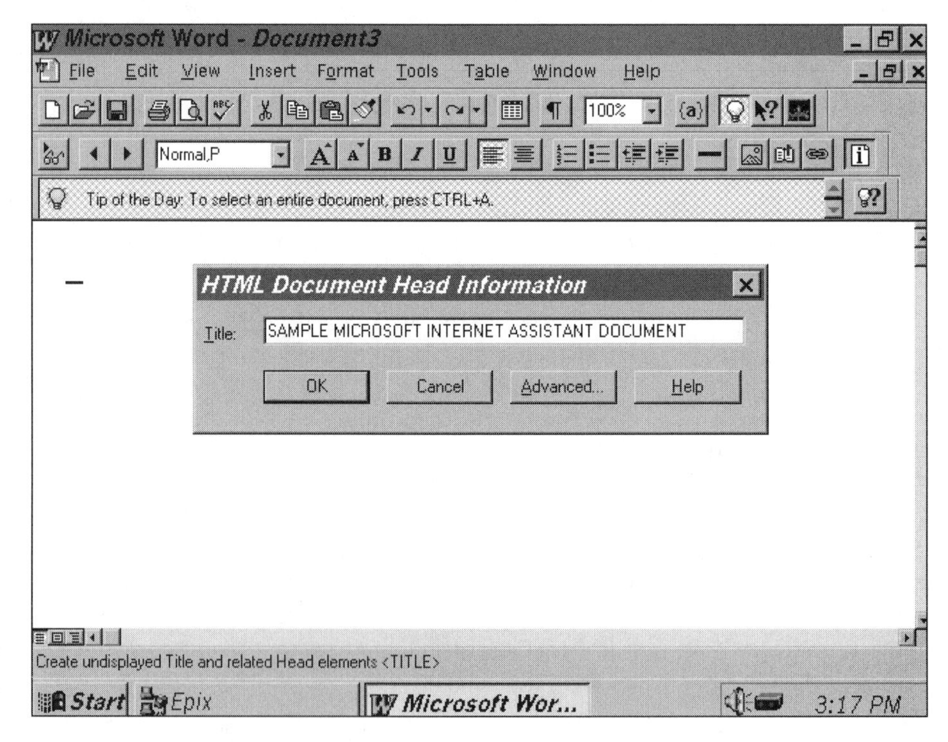

Figure 3-10.
The HTML Document Head Information Dialog Box in Microsoft Word

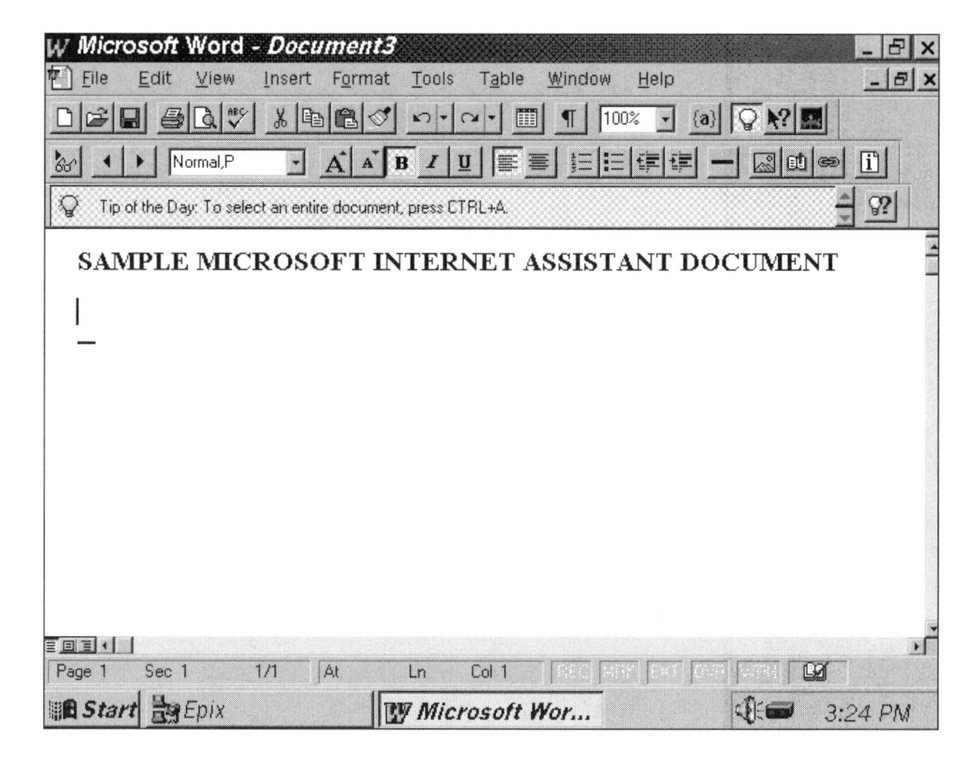

Figure 3-11. HTML Document with Title in Microsoft Internet Assistant

9. Click on Hyperlink on the Insert menu. Key http://www.microsoft.com/msword in the File or URL typing box in the Hyperlink dialog box (see Illustration 3-12). Click on OK (see Figure 3-13). Click on Save on the File menu.
10. If you are presently online to the Internet, double-click on the blue URL. A series of downloading boxes will appear. The Web page will appear in the Microsoft Internet Assistant Browser (see Figure 3-14).

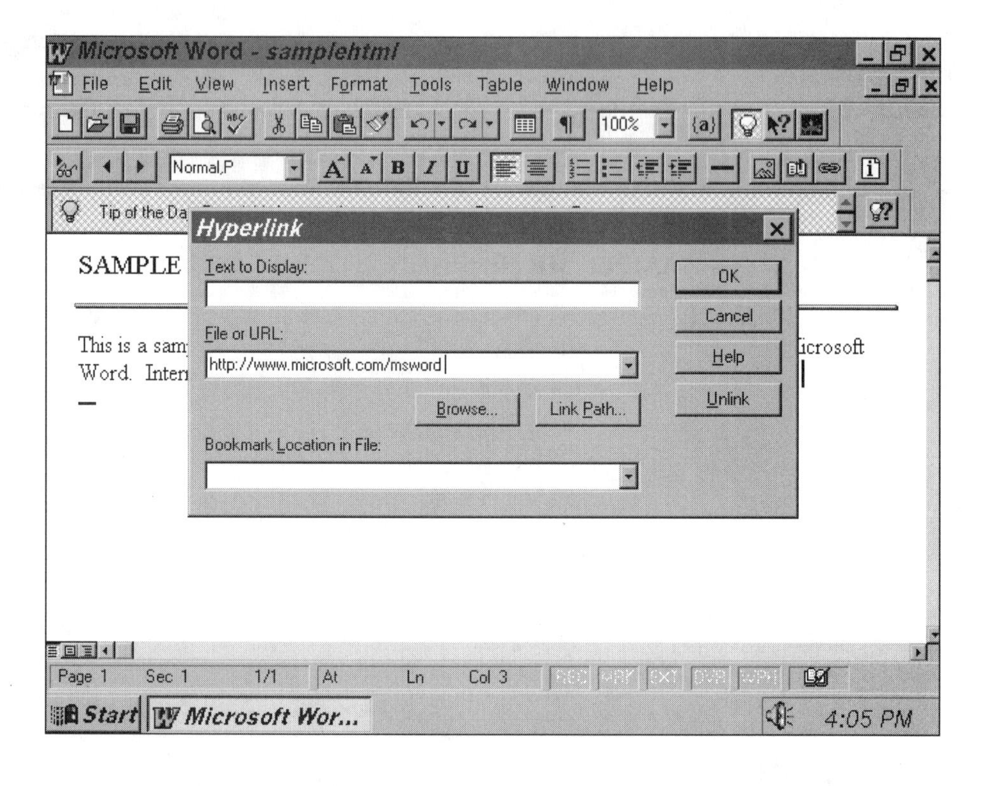

Figure 3-12. Hyperlink Dialog Box in Microsoft Internet Assistant

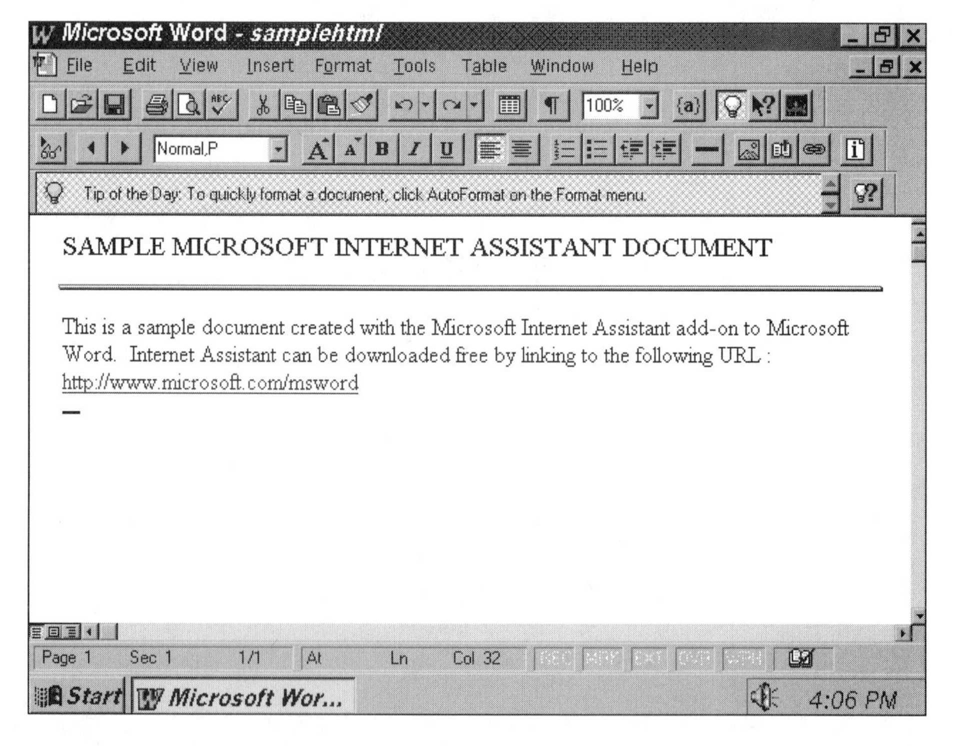

Figure 3-13. HTML Document with Hyperlink in Microsoft Internet Assistant

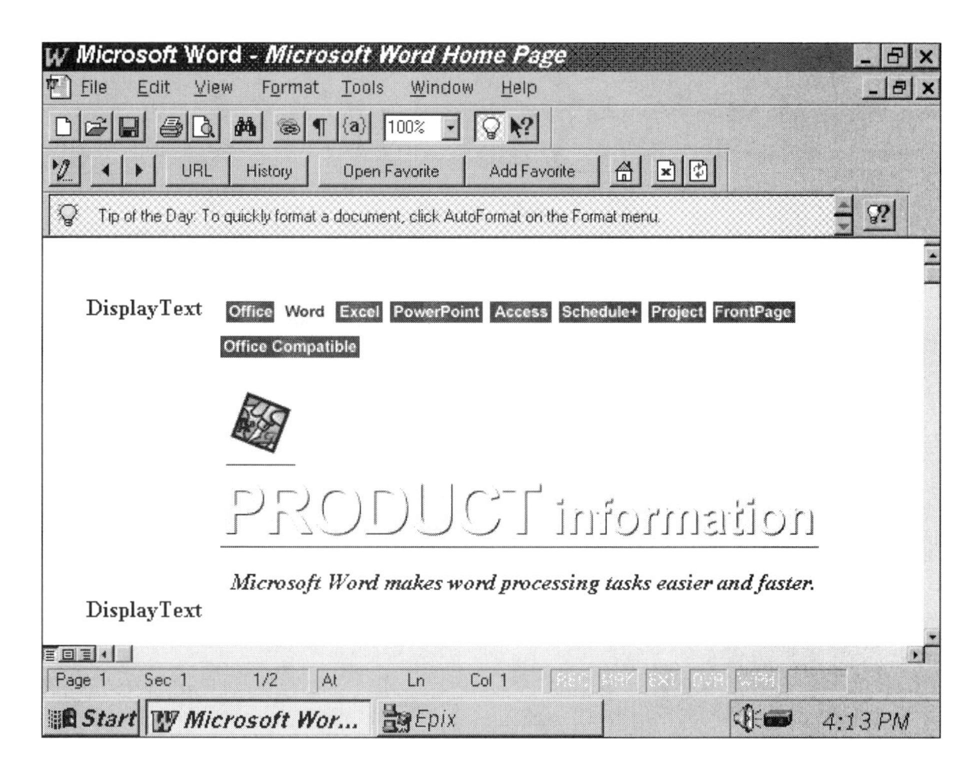

Figure 3-14. Linked Web Page in Microsoft Internet Assistant Browser

Steps and Procedures for Using WordPerfect Internet Publisher

To use Internet Publisher you must have a copy of Netscape and WordPerfect for Windows 6.1 installed on your computer. After downloading Internet Publisher, you must install the add-on program. Run the self-executing wpipzip.exe file and follow the onscreen instructions. You will be prompted for the paths of WordPerfect and Netscape.

To use Internet Publisher to create a simple HTML file, follow these steps:

1. Run WordPerfect.
2. Click on New on the File menu or on the New Document button on the toolbar. (The New Document dialog box should appear.)
3. Click on HTML document under Select Template.
4. Click on the Select button. (A new document screen and the HTML toolbar should appear; see Figure 3-15).
5. Key Internet Publisher as the document title in the Document Title dialog box and click on OK.
6. Key in the following paragraphs, pressing the Enter key twice after the first paragraph.

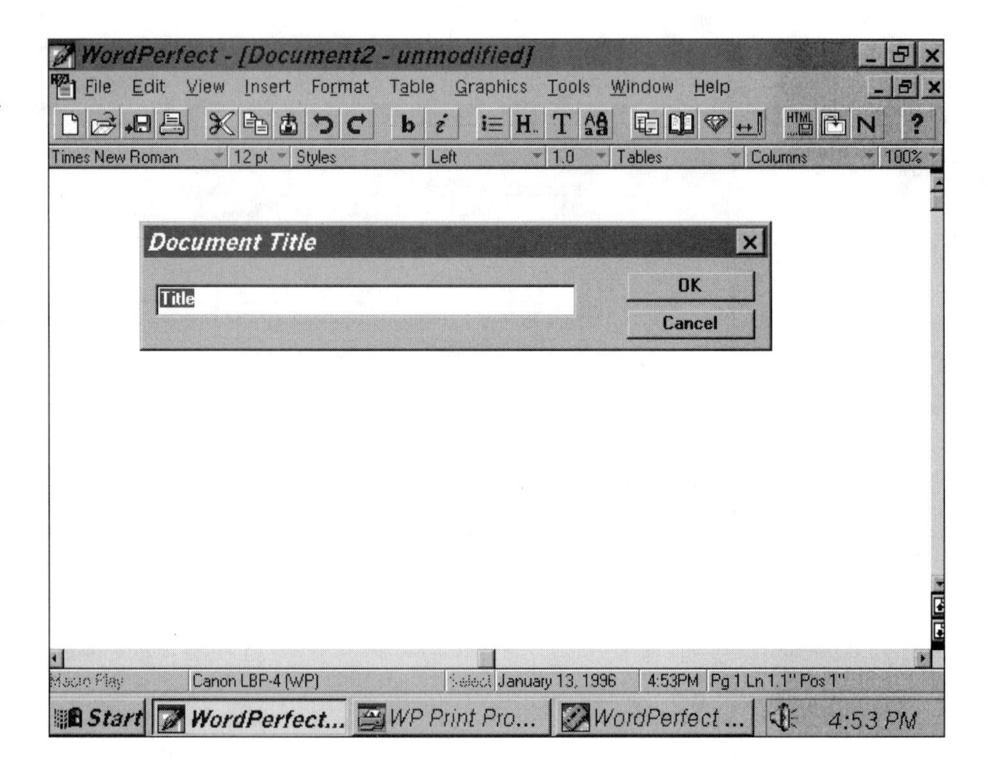

Figure 3-15. WordPerfect with Internet Publisher Loaded

Welcome to Internet Publisher. This easy-to-use software tool for WordPerfect 6.1 for Windows allows you to create HyperText Markup Language (HTML) documents for publishing on the World Wide Web (WWW). You can also use the included Netscape Navigator browser for viewing pages (or documents) on the WWW.

The Internet Publisher also comes with an integrated Envoy document viewer. This viewer allows you to read any Envoy document on the WWW in its native format (.EVY).

7. Click on Horizontal Line on the HTML menu or on the Horizontal Line button on the toolbar and press the Enter key.
8. Key Internet Publisher offers the following advantages: and press the Enter key.
9. Click on List on the HTML menu or on the List button on the toolbar.
10. Click on Button List (UL) to select it and click on OK.
11. Key You don't need to learn the format for the HTML tags. next to the bullet and press the Enter key.
12. Key You can use WordPerfect for Windows, which you may be already familiar with, next to the bullet and press the Backspace key to delete the next bullet.

13. Click on Heading on the HTML menu or on the Heading button on the toolbar.
14. Click to select Address under Other Paragraph Types and click on OK.
15. Key (Your name); Enter.
    Key (Your address).
16. With a diskette in drive A, click on Save As... on the File menu; key A:intpub as the file menu; and click on OK.
17. Click on Export as HTML on the File menu or click on the Export HTML button on the toolbar. (Netscape should load and the document should appear. This is how the file will appear in the browser.)
18. Click on Exit on the File menu of Netscape to return to WordPerfect.
19. Click on Open on the File menu or on the Open button on the toolbar.
20. Open the intpub.htm file, which should be on your A disk.
21. Click on the down scroll arrow to the right of Unknown Format and scroll to ASCII (DOS) Text and click on it. Click on OK. (The created HTML file should appear; see Figure 3-16.)

Figure 3-16. WordPerfect with Created HTML File

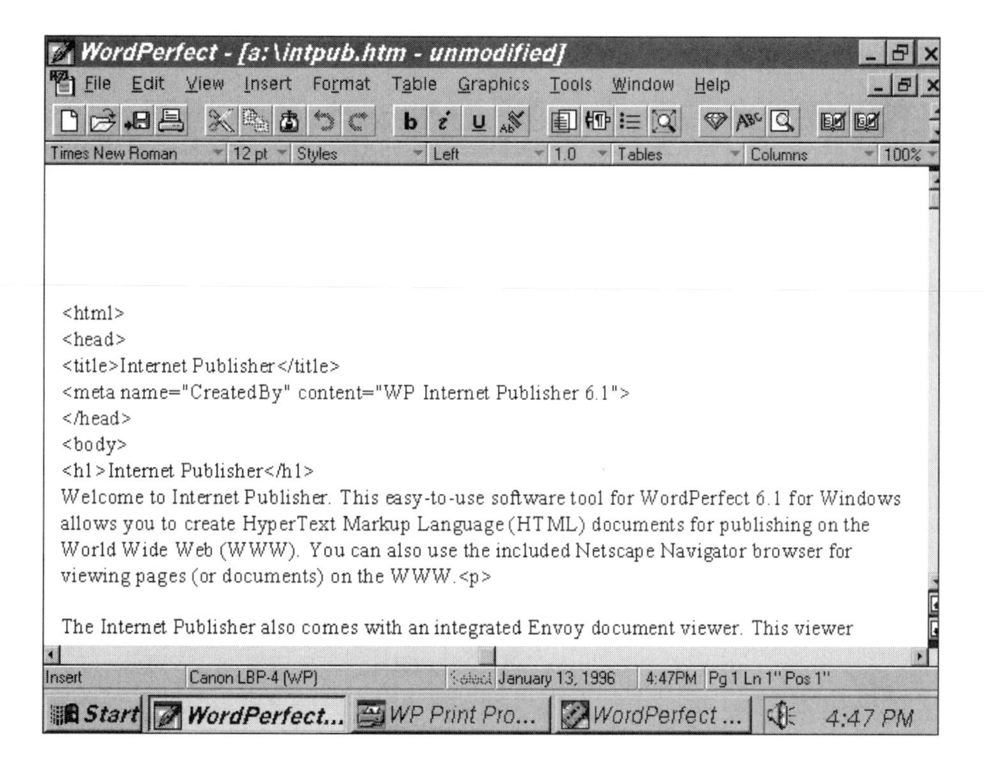

You can also use Internet Publisher to add hyperlinks by choosing Hypertext Link from the HTML menu or by clicking on the Hypertext Link button on the toolbar. Graphics can be added to the document by choosing Graphic from the HTML menu or by clicking on the Graphic button on the toolbar. In each case, dialog boxes appear to assist you in keying the URLs to the link.

What do you do if you don't have Microsoft Word or WordPerfect available to use an add-on editor? The answer is to use a standalone HTML editor, which is available for both IBM and IBM PC–compatible and Macintosh computers. One of the differences in using a standalone HTML editor is that you definitely need to have some knowledge of HTML tags. If you skipped over the previous section on HTML tags, now is the time to go back and read it.

IBM AND IBM PC–COMPATIBLE STANDALONE HTML EDITORS

We will examine and briefly explain the following HTML editors for IBM and IBM PC–compatible computers: HTML HyperEdit, HTML Assistant, HoTMetaL, and HTML Writer. We will provide a more detailed explanation of how to use the HTML Writer program by showing how to create a simple HTML file. Although other free or inexpensive editors are available, these four seem to be most frequently used by those users who write HTML documents on IBM PC–compatible computers.

HTML HYPEREDIT

The HTML HyperEdit was written by Stephen Hancock (e-mail: s.hancock@info.curtin.edu.au). HyperEdit contains some interesting features, such as Search and Replace and a tutorial, that other editors may not have. Missing, however, is any test feature to allow you to go to a browser like Netscape from within the editor; to see how the HTML file will actually be displayed, you must exit the program first. Both the HTML Assistant and HTML Writer editors have a built-in test feature.

HTML HyperEdit has a beginner's mode and advanced modes. The beginner's mode has fewer buttons to simplify the screen. The advanced mode adds the following buttons: Images, Custom Formatting Tags, Ampersand and Bracket Conversion, Signatures/Footers, and Shortcut keys. The Signature/Footers button allows you to add lines to the HTML file to identify the person, department, and other information related to the document.

The HTML HyperEdit editor is available by downloading the htmledit.zip file by using a download utility, Netscape, or another browser at ftp.sunet.se with the path /pub/www/utilities/hyperedit (ftp://ftp.sunet.se/pub/www/utilities/hyperedit). After unzipping the file, place the resulting program files in the subdirectory C:\HTMLEDIT. To install the program, key in the following for the Windows command line: C:\HTMLEDIT\TBOOK HTMLEDIT.TBK. You can substitute another drive letter, if necessary. Click on the resulting icon to start the program. Figure 3-17 shows the HTML HyperEdit editor.

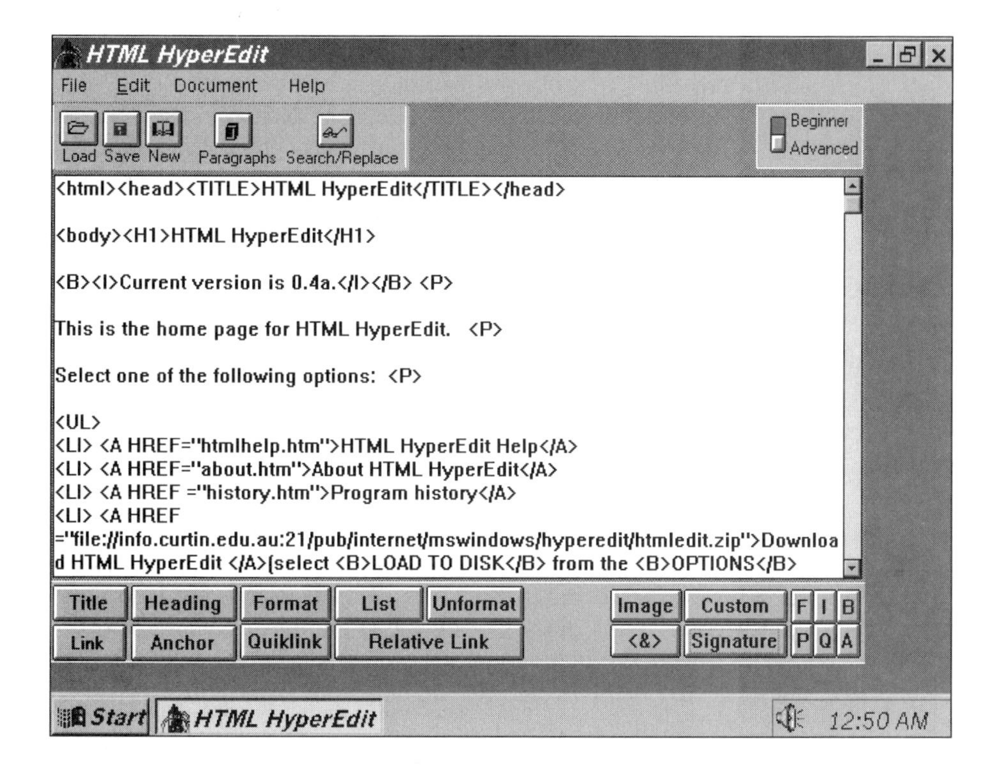

Figure 3-17. The HTML HyperEdit Editor

HTML ASSISTANT

The HTML Assistant, written by Howard Harawitz (e-mail: harawitz@fox.nstn.ns.ca), is the freeware version of HTML Assistant Pro. The commercial version adds the following features: an automatic page creator and options to remove HTML markings from text, to make UNIX text files more readable by converting them to DOS text, and to save DOS text files as UNIX text. HTML Assistant Pro also makes enhancements and upgrades available.

The freeware version, however, is a fully functional editor with many well-developed features. An extensive number of buttons at the top of the screen cover most of the HTML tags, and an explanation of each of the buttons is displayed on the status line at the bottom of the screen when you place the mouse pointer on each button. The following buttons are available when you retrieve an existing file or create a new one: Undo, Repeat Markup, Recall, Hold, Save, and Test. The Undo button allows you to erase the last tags you entered in case you change your mind. The Repeat Markup button provides a quick way of entering the last tag without needing to find the original command button. The Test button provides a way of displaying the created pages in a browser. For the button to be functional, you need to choose Enter Test Program Name from the File menu. The program will ask you to indicate the path for your browser.

The HTML Assistant is available by downloading the htmlasst.zip file by using a download utility or Netscape at the following addresses: ftp.cs.dal.ca with the path /htmlasst (ftp://ftp.cs.dal.ca/htmlasst) or ftp.nau.edu with the path /pub/windows/html_asst (ftp://ftp.nau.edu/pub/windows/html_asst). After unzipping the file, the resulting program files should be placed in the subdirectory C:\HTMLASST. The following files should be placed in the C:\WINDOWS\SYSTEM subdirectory: cmdialog.vbx and threed.vbx. You may also need to download the file vbrun300.zip at the first address above or from some other server on the Internet and also place the resulting vbrun300.dll in your C:\WINDOWS\SYSTEM subdirectory. Figure 3-18 shows the HTML Assistant editor. You can substitute another drive letter, if necessary. Create an icon for the program and click on the icon to start the editor.

HOTMETAL

The HoTMetaL editor is a rule-based HTML editor written for both IBM or IBM PC–compatible and Macintosh computers. HoTMetaL's most important features are rules checking and validation, which can prevent you from creating invalid HTML markup. *Rules checking* prevents markup errors as you're editing, and *validation* ensures that the markup is correct and complete. HotMetaL's unique feature is its display of tags as a graphical element. HotMetaL also provides a convenient table model that Web browsers can display. The shareware version of HoTMetaL is available at http://www.sq.com. Figure 3-19 shows the HoTMetaL editor.

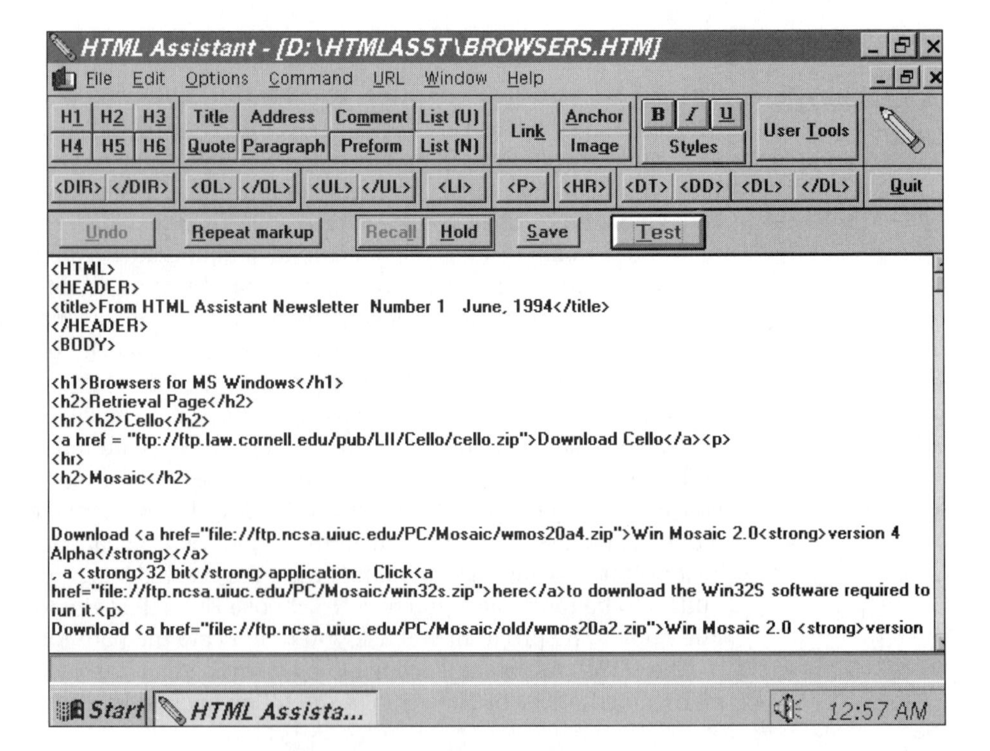

Figure 3-18. The HTML Assistant Editor

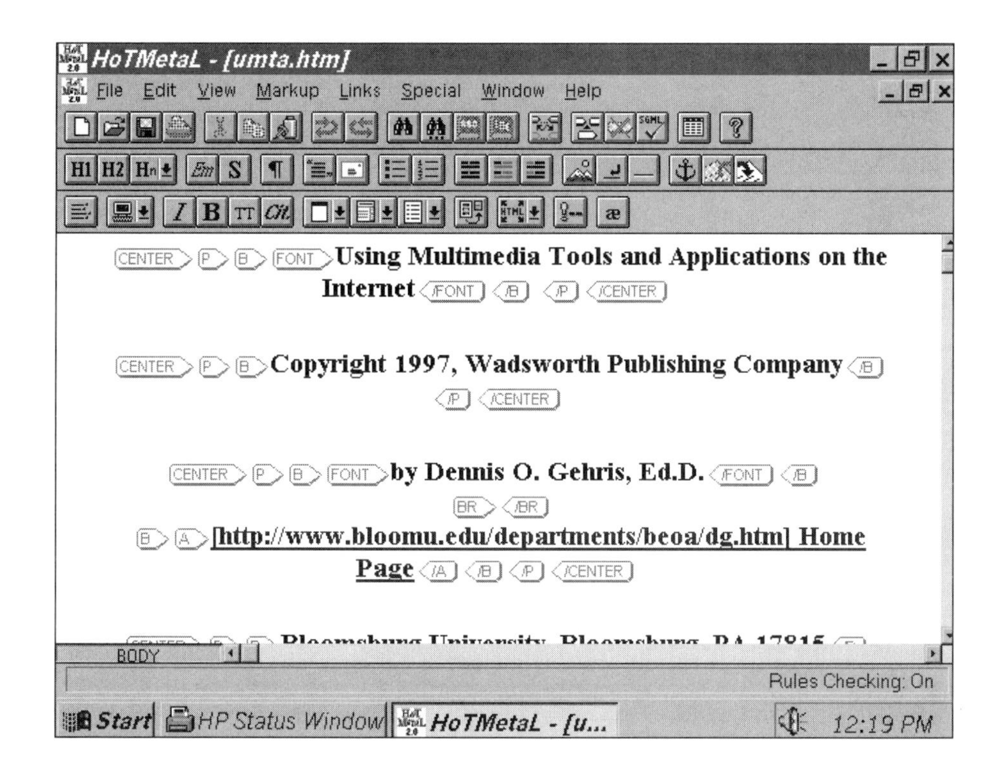

Figure 3-19. The HoTMetaL Editor

HTML WRITER

This HTML editor is the author's personal favorite and he recommends it highly. It works under both the Windows 3.1 and Windows 95 operating systems. The HTML Writer was written by Chris Nosack (e-mail: html-writer@byu.edu or WWW: http://lal.cs.byu.edu/people/nosack/), who asks that a donation of $10 be sent to him after 30 days at 376 North Main Street, Orem, Utah 84057. HTML Writer supports every HTML tag, and most of the tags are represented by a button on the button bar at the top of the screen. An explanation of each button is displayed in a rectangle near the button and on the status line at the bottom of the screen when you place the mouse pointer on each button. All the tags are accessible through the HTML menu that is displayed along with the File, Edit, Search, Options, Window, and Help menus after you open an existing HTML file or when you create a new one. A good help system is available that should answer any question you may have. The HTML Writer also includes a Test feature that permits you to view your HTML pages in the browser of your choice. You will be prompted for the browser path and file if you click on the Test button or on Test... in the File menu without previously configuring the browser.

The HTMLWriter editor is available by using one of the following URLs with Netscape or another browser: http://lal.cs.byu.edu/people/nosack/ or ftp://ftp.coast.net/SimTel/win3/internet/. The filename should begin with hw, but the complete filename will vary depending on the version of the editor. After unzipping the

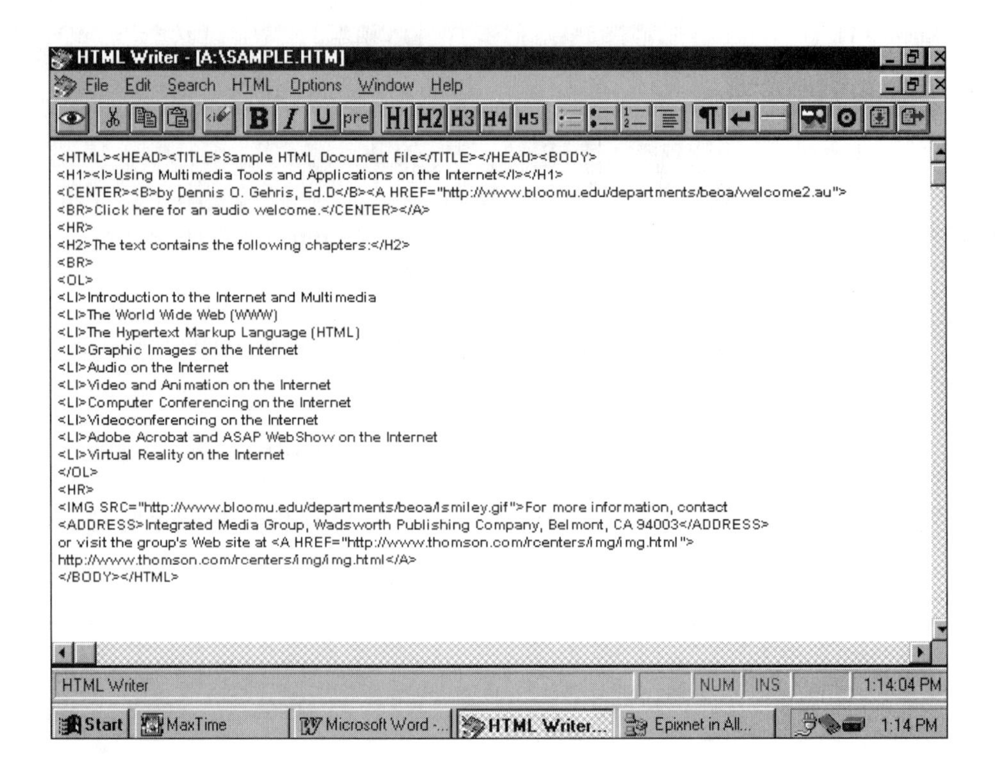

Figure 3-20.
The HTML Writer Editor

file, place the resulting program files in the subdirectory C:\HTMLWRIT. The following files should be placed in the C:\WINDOWS\SYSTEM subdirectory: cmdialog.vbx, commdlg.dll, toolbars.vbx, and emedit.vbx. You may also need to download the file vbrun300.zip and also place the resulting vbrun200.dll in your C:\WINDOWS\SYSTEM subdirectory. You can substitute another drive letter, if necessary. Create an icon for the program and click on the icon to start the editor. Figure 3-20 shows the HTML Writer editor.

Steps and Procedures for Using HTML Writer

The HTML Writer editor is a simple program to use. To create a short HTML file with a title, unordered list, and address, follow these steps:

1. Launch the editor by clicking on the HTML Writer icon.
2. Click on New on the File menu to open a new file. (A new file window should appear.) Maximize the window.
3. Click on Upper Case Tags on the Options menu.

4. Click on the Bold font style on the Options/Screen Font menu. Click on OK.
5. Click on HTML on the HTML/Document menu. (HTML tags should appear.)
6. Click on Head on the HTML/Document menu. (Head tags should appear.)
7. Key PC HTML Editors. Use your mouse to drag the mouse cursor over these words to select them.
8. Click on Title on the HTML/Document menu. (Title tags should appear.)
9. Click an insertion point between the </HEAD> and </HTML> tags. Press the Enter key.
10. Click on Body on the HTML/Document menu. (Body tags should appear.) Press the Enter key.
11. Click an insertion point before the </BODY> tag if necessary.
12. Key PC HTML Editors. Use your mouse to select these words.
13. Click on the H1 button. (H1 codes should appear.)
14. Click an insertion point between the </H1> and </BODY> tags. Press the Enter key.
15. Key the following:

    HTML HyperEdit (Enter)
    HTML Assistant (Enter)
    HoTMetaL (Enter)
    HTML Writer (Enter)
16. Use your mouse to select these lines.
17. Click on the Bulleted List button. (List tags should appear.)
18. Key the following:

    (Your Name) Click on the Line break button. (
 should appear.)
    (Enter) (Your Address)
19. Use your mouse to select these lines.
20. Select Address from the HTML/Style menu. Compare your screen with Figure 3-21.
21. Select Save As from the File menu. Select the path and save the file with the filename sample1.htm.
22. Click on the Test button. You may be asked to configure your browser. Compare your screen with Figure 3-22.

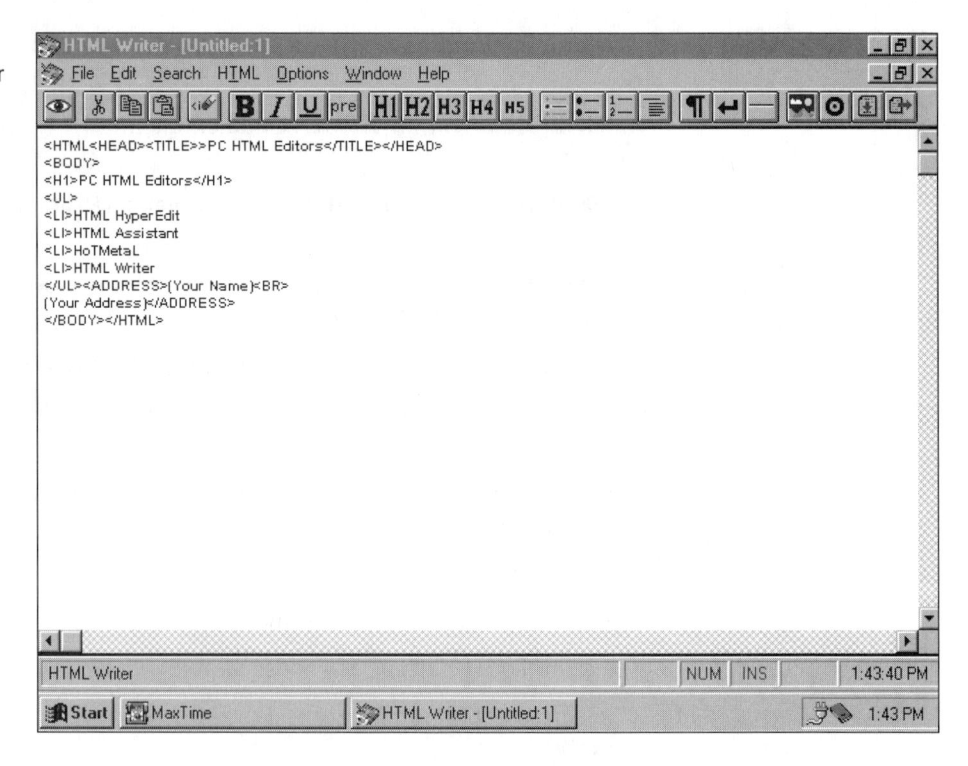

Figure 3-21. HTML Writer Editor with Completed Steps

23. Print the document if a printer is available.
24. Exit the browser.
25. Select Exit from the File menu to exit HTML Writer.

To insert a graphic image into your file, click on the Inline image button. The Inline image dialog box should appear (see Figure 3-23). Enter the URL in the box or click the build button to obtain assistance on entering the URL. Click on OK and the tag will be inserted.

To insert a hyperlink into your file, click on either the Local hyperlink button or the Remote hyperlink button. In each case, a dialog box will appear into which you enter the correct URL. Clicking on OK will enter the tags into your file. For local hyperlinks, select a word in an existing file, then click on the Target button and you will be prompted for a target name. These target names are then specified in the Local Hyperlink Dialog box.

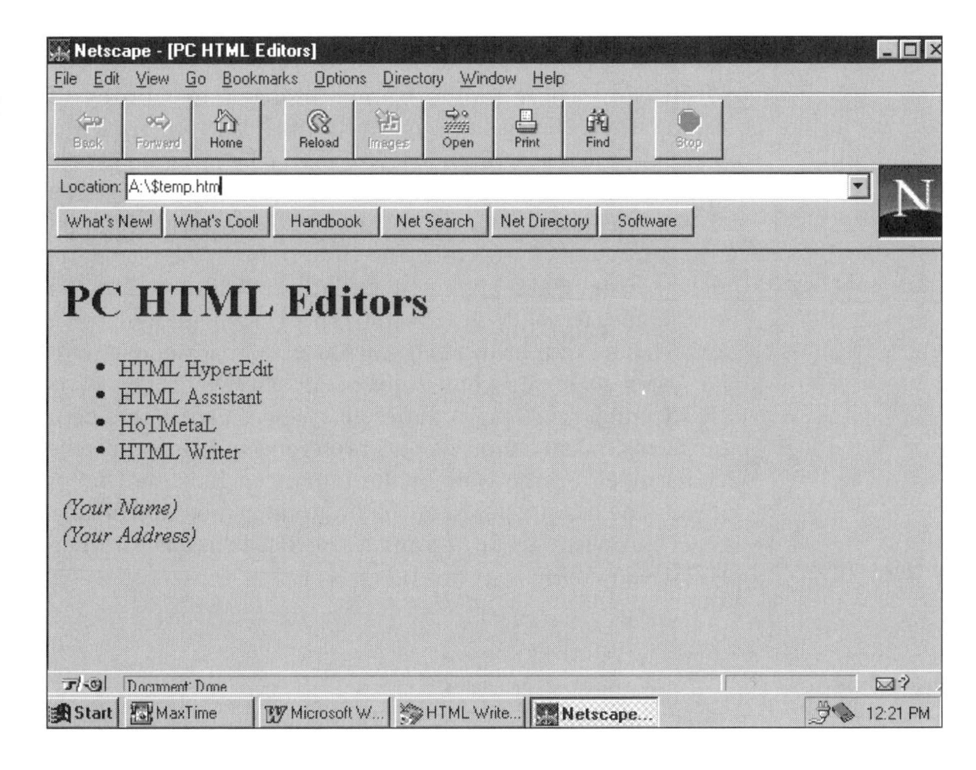

Figure 3-22. Netscape Browser with Test of Completed Steps

Figure 3-23. The HTML Writer Inline Image Dialog Box

MACINTOSH HTML EDITORS

Three HTML editors for Macintosh computers will be examined and briefly explained: HTML.edit, Arachnid, and HTML Pro. Although other free or inexpensive editors are available, these are the three that seem to be most frequently used by those who write HTML documents on Macintosh computers. A more detailed explanation of how to use the HTML Pro program will be provided by showing how to create a simple HTML file.

HTML.EDIT

HTML.edit, a freeware program designed by Murray M. Altheim (e-mail: murray.altheim@nttc.edu), requires 2 MB RAM (prefers 4 MB), is limited to 32K text files, and was written in HyperCard. The program uses a window that is easily resizable and an interface that presents tag options in a very nice manner. HTML.edit supports tables, forms, and other tags. A built-in hypertext help system is available by clicking the Help button (the question mark) in the upper right corner of the application window. You can use the Preview menu (the world icon) to select the Preview This Document item to view the current document using your browser. If you haven't already configured a browser, you may be asked to locate your copy of Mosaic, Netscape, or other Web browser. HTML.edit possesses five different types of cards (screens): index, editor, master heads, master footers, and preferences. You begin editing a document by choosing either the New or the Import an Existing One option from the File menu. The latest version of the program can be downloaded with Netscape or other browser at ftp://ftp.tidbits.com/tidbits/tisk/html. Figure 3-24 shows the HTML.edit editor.

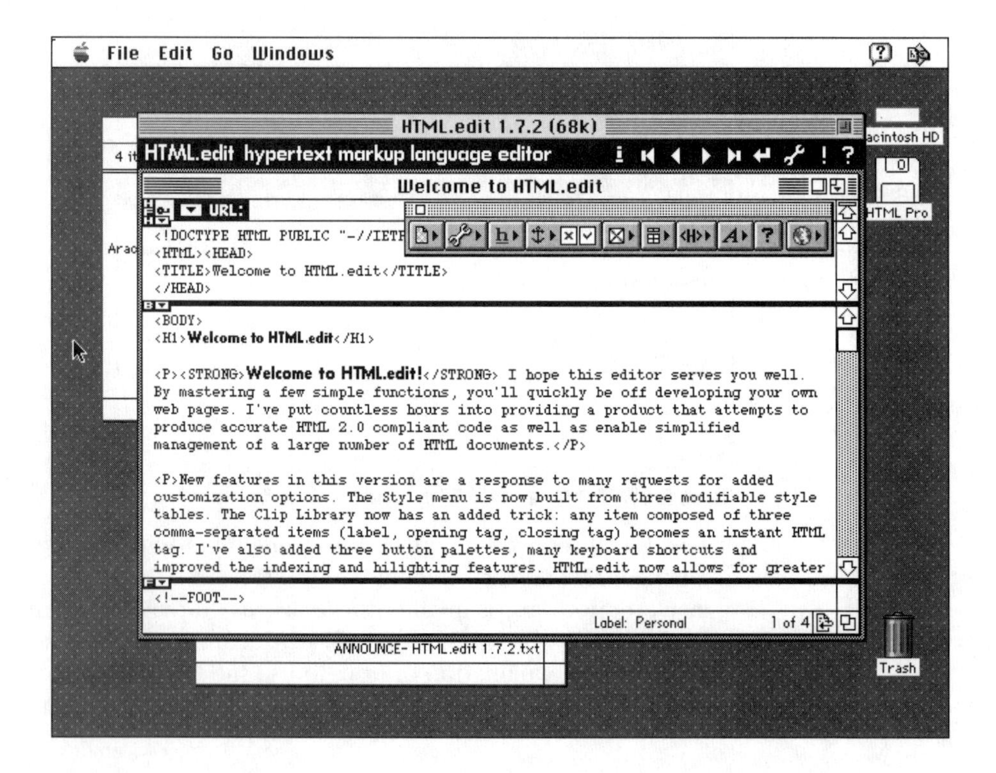

Figure 3-24. The HTML.edit Editor

ARACHNID

Arachnid, another freeware editor, was written by Robert McBurney at the University of Iowa (e-mail: robert-mcburney@uiowa.edu). It requires 2 MB RAM (prefers 4 MB), is limited to 32K text files, has a WYSIWYG page view, and was written in SuperCard. The program has an interesting form creation interface and allows for the display of inline images. The latest version of the program can be downloaded with Netscape or other browser at ftp://ftp.mc.hik.se/pub /mac/html/archnid/. Figure 3-25 shows the Arachnid editor.

HTML PRO

HTML Pro, a shareware program with a $5 registration fee, was written by Niklas Frykholm in Sweden (e-mail: nisfrm95@student.umn.se). It can be run in 500K RAM, is limited to 32K text files, and has a WYSIWYG page view. HTML Pro is a very good editor for beginners because editing is done in two windows. One window provides a somewhat what-you-see-is-what-you-get view of the document and the other the HTML coded text. Keying HTML codes can be avoided by using the tag listing on the Styles menu. One of the limitations of the program is that you can only display one document at a time. It is also difficult to change from one tag to another and there are no features for multidocument management.

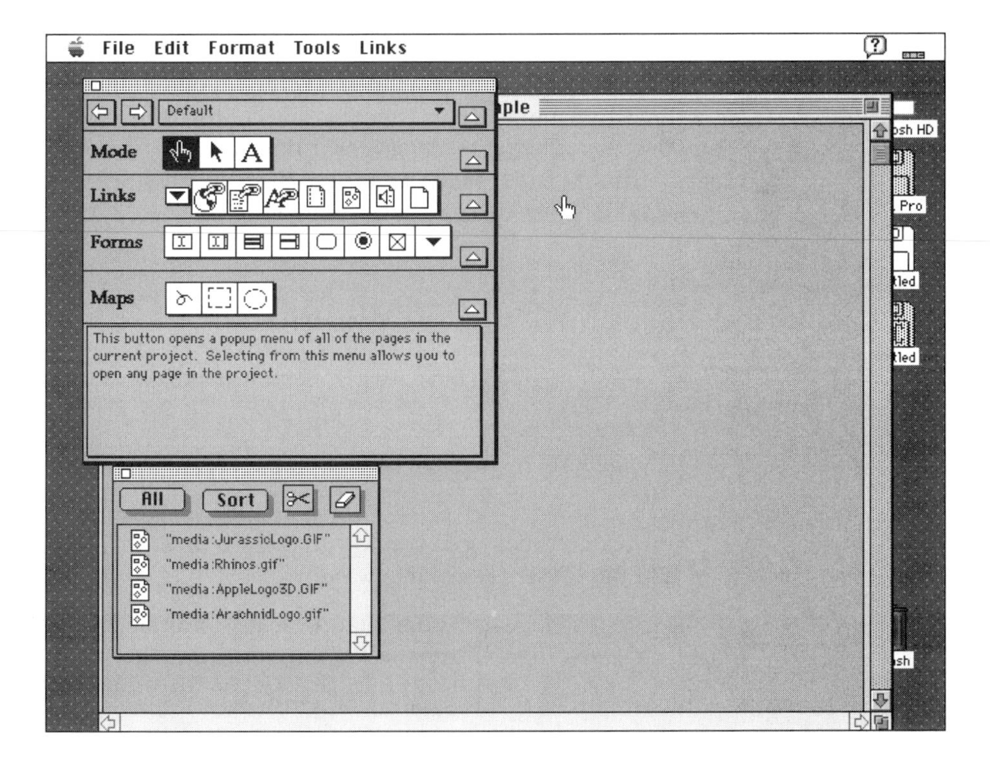

Figure 3-25. The Arachnid Editor

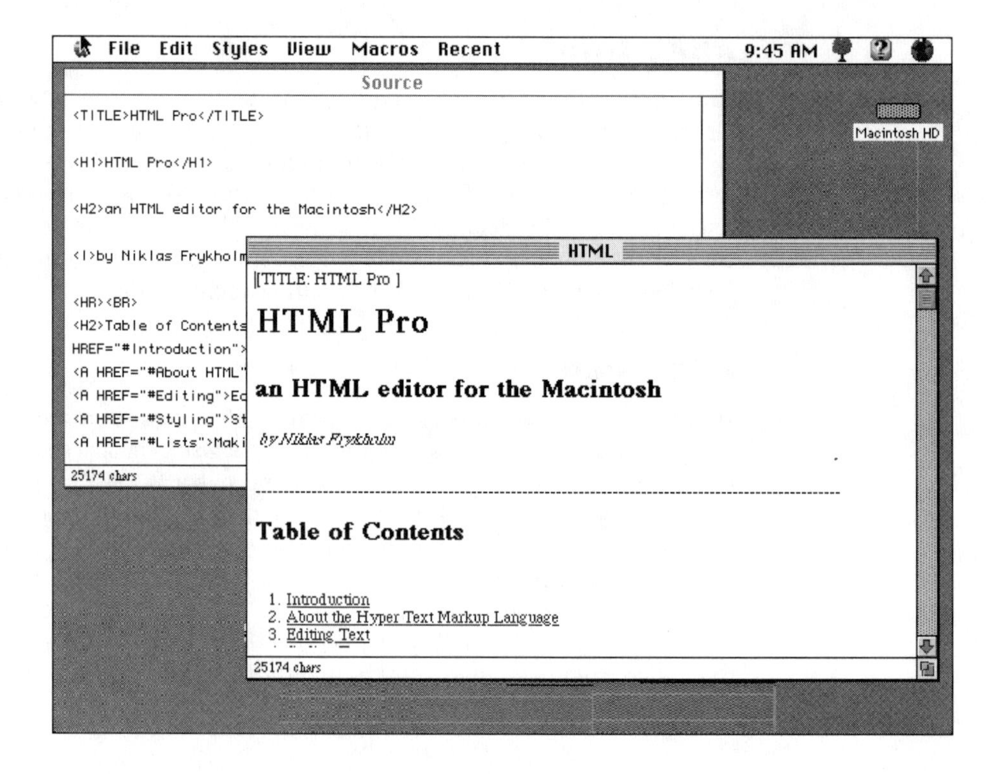

Figure 3-26. The HTML Pro Editor

The HTML Pro editor is available by downloading the htmpro1.08.cpt.hqx file by using a download utility, Netscape, or another browser at ftp.luth.se with the path /pub/mac/.2/mirror.umich/util/comm (ftp://ftp.luth.se/pub/mac/.2/mirror.umich/util/comm). The filename may change as later versions are released. Figure 3-26 shows a HTML Pro editor.

Steps and Procedures for Using HTML Pro

To create a short HTML file with a title, unordered list, and address, follow these steps:

1. Launch the editor by clicking on the HTML Pro application program icon.
2. Click on the Source window to bring it in view. You will be placing all of the HTML tags in this window.
3. Key <HTML> tag. Notice that pressing < produces <> with the cursor between them. Press the Right Arrow key.
4. Key <HEAD>. Press the Right Arrow key.

5. Key <TITLE>. Press the Right Arrow key.
6. Key Macintosh HTML Editors.
7. Key </TITLE>. Press the Right Arrow key.
8. Key </HEAD>. Press the Right Arrow key. Press the Return key.
9. Key <BODY>. Press the Right Arrow key.
10. Key Macintosh HTML Editors. Use your mouse to drag the mouse cursor over these words to select them.
11. Click on Header 1 on the Styles/Headers menu. (H1 codes should appear.)
12. Click an insertion point after </H1>. Press the Return key.
13. Key the following:

    HTML.edit (Return)
    Arachnid (Return)
    HTML Pro (Return)
14. Use your mouse to select these words.
15. Click on Unordered Lists on the Styles/Lists menu. (List tags should appear.)
16. Add <LI> codes before the Arachnid and HTML Pro lines.
17. Click an insertion point after </UL>. Press the Return key.
18. Key the following:

    (Your Name)
 (Return)
    (Your Address)
19. Use your mouse to select these lines.
20. Click on Address on the Styles/Logical menu. (Address tags should appear.)
21. Click an insertion point after </ADDRESS>/.
22. Key </BODY></HTML>.
23. Click on Save As on the File menu. Select the path and save the file with the filename Sample2.
24. Click on HTML on the View menu. Compare your screen with Figure 3-27. You may also want to view a document in your browser. Click on Launch Browser on the View menu.
25. Select Quit from the File menu to exit HTML Pro.

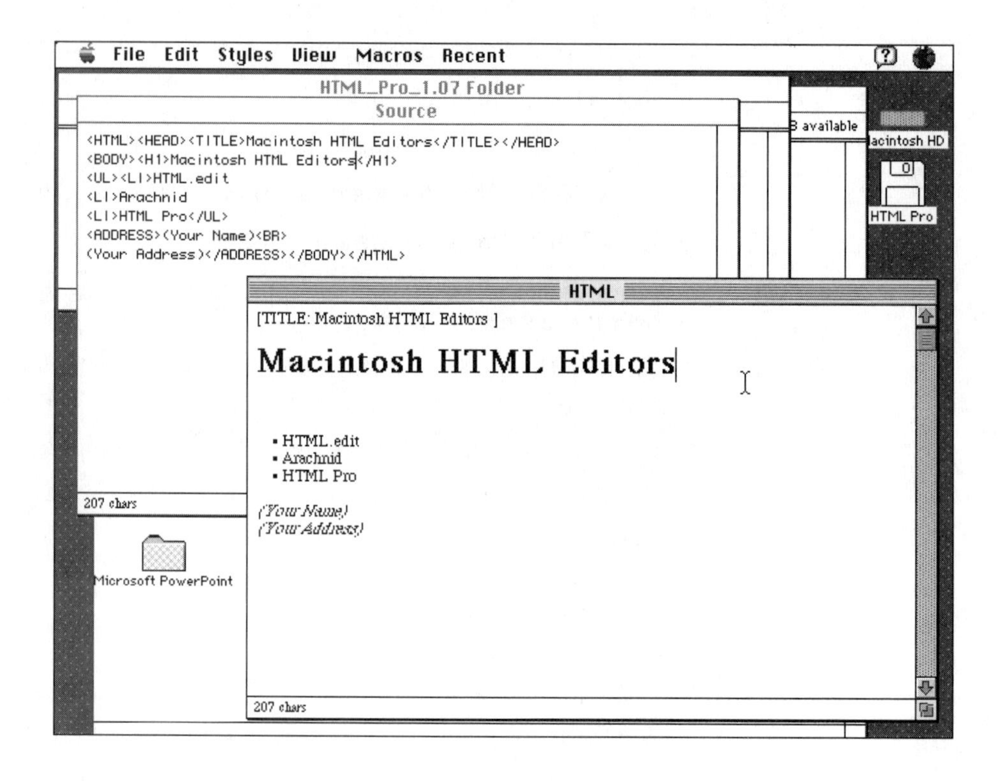

Figure 3-27. HTML Pro Editor with Completed Steps

Questions for Review

1. Define HTML. What is HTML's purpose?
2. What are the advantages of HTML?
3. What are the disadvantages of HTML?
4. Identify the functions of each of the following types of HTML tags: structural, paragraph formatting, character formatting, list specification, hyperlink, and multimedia.
5. What is the purpose of HTML software utility programs? What is a converter program? What is an editor program?

Exercises

1. Use Netscape or another browser to open the following documents at the URLs indicated. Study each document.

   Beginner's Guide to HTML
   http://www.ncsa.uiuc.edu/General/Internet/WWW/HTMLPrimer.html

   Beginner's Guide to URLs
   http://www.ncsa.uiuc.edu/demoweb/url-primer.html

2. Use one of the HTML editor programs to produce a file with the tags as shown below. Save the file with the filename exer2.htm (exercise2 if you are using a Macintosh computer). Use your browser to view the document. Print the page from your browser.

```
<HTML><HEAD><TITLE>A Glossary of Internet Terms</TITLE></HEAD>
<BODY><H1>Glossary of Internet Terms</H1>
<HR>
<H2>The following is a glossary of important terms related to the Internet.</H2>
<DL>
<DT><B>Internet</B>
<DD>A network of computer systems that are interconnected in about 130
countries.
<HR>
<DT><B>World Wide Web</B>
<DD>What makes multimedia applications possible on the Internet.
<HR>
<DT><B>FTP</B>
<DD>This is "file transfer protocol" and is used to exchange files on the
Internet.
<HR>
<DT><B>Gopher</B>
<DD>An Internet service that consists of a system of menus.
<HR>
<DT><B>Telnet</B>
<DD>This allows the user to log on to remote computers connected to the
Internet to run programs or access information.</DL>
<HR>
<ADDRESS>(Your Name)<BR>
(Your Address)</ADDRESS>
</BODY></HTML>
```

3. Use one of the HTML editor programs to produce a file with the tags as shown below. Save the file with the filename exer3.htm (exercise3 if you are using a Macintosh computer). Use your browser to view the document. Print the page from your browser.

```
<HTML><HEAD><TITLE>HTML Editors for Microsoft Windows</TITLE></HEAD>
<BODY>
<H1>HTML Editors for Microsoft Windows</H1>
<H2>Retrieval Page</H2>
<HR>
<P><H3>This page provides the addresses for the following HTML Editors:</H3>
<UL><LI>HTML HyperEdit
<LI>HTML Assistant
<LI>HoTMetaL
<LI>HTML Writer</UL>
<HR>
<P>HTML HyperEdit can be downloaded by clicking on <A
HREF="ftp://ftp.sunet.se/pub/www/utilities/hyperedit/hyperedit.zip">download
</A>.
<HR>
<P>HTML Assistant can be downloaded by clicking on <A
HREF="ftp://ftp.cs.dal.ca/htmlasst/htmlasst.zip">download</A>.
<HR>
<P>HoTMetaL can be downloaded by clicking on <A
HREF="http://www.sq.com">download</A>.
<HR>
<P>HTML Writer can be downloaded by clicking on <A
HREF="ftp://lal..cs.byu.edu/pub/www/tools/htmlwrit.zip">download</A>.
<HR>
This page was prepared by
<P><ADDRESS>(Your Name)<BR>
(Your Address)</ADDRESS></BODY></HTML>
```

4. Use one of the HTML editor programs to produce a file with the tags as shown below. Save the file with the filename exer4.htm (exercise4 if you are using a Macintosh computer). Use your browser to view the document. Print the page from your browser.

```
<HTML><HEAD><TITLE>HTML Editors for the Macintosh</TITLE></HEAD>
<BODY>
<H1>HTML Editors for the Macintosh</H1>
<H2>Retrieval Page</H2>
<HR>
<P><H3>This page provides the addresses for the following HTML Editors:</H3>
<UL><LI>HTML.edit
```

```
<LI>Arachnid
<LI>HTML Pro</UL>
<HR>
<P>Information about downloading HTML.edit can be obtained by clicking on
<A
HREF="http://www.ogopogo.ntcc.edu/tools/HTMLedit/HTMLedit.html">Informa-
tion on HTML.edit</A>.
<HR>
<P>Information about downloading Arachnid can be obtained by clicking on
<A HREF="ftp://ftp.mc.hik.se/pub/mac/html/arachnid">Information on
Arachnid</A>.
<HR>
<P>HTML Pro can be downloaded by clicking on
<A
HREF="ftp://ftp.luth.se/pub/mac/.2/mirror.umich/util/comm/htmpro1.08.cpt.hqx">
Download HTML Pro</A>.
<HR>
This page was prepared by
<P><ADDRESS>(Your Name)<BR>
(Your Address)</ADDRESS></BODY></HTML>
```

5. Identify and explain the purpose of each HTML tag in the sample HTML document file in the chapter. Key and save the file (filename: sample.htm or sample.html), using one of the HTML editor programs. If you are connected to the Internet, the “smiley” graphic should appear. Click on Click Here for An Audio Welcome to play the sound file.

CHAPTER

4

Graphic Images on the Internet

This chapter will explain how to use graphics with the World Wide Web. It will define graphic image, discuss the graphic image file formats, show how to use graphic images, and provide step-by-step instructions for obtaining and creating graphic files.

What You Will Learn

- How a graphic image is defined
- Graphic image file formats
- How to use graphic images
- How to obtain graphic image files
- How to find and download graphic images on the Internet
- How to create graphic image files with a scanner
- How to create graphic image files using video capture

Graphic Image Defined

As we saw in Chapter 1, multimedia is a combination of text, graphics, audio, and full-motion video. The most frequent element after text is the graphic image. Anyone familiar with an Internet system based solely on text (i.e., UNIX) can testify that the material becomes very boring. Graphic images are what sets a text-based system apart from a graphical-based system and helps to maintain the interest of the person viewing a Web page. A **graphic image** can consist of a drawing or a static (demonstrating no movement) photograph.

Graphic images on a Web page may include photographs, drawings, paintings, charts, graphics, tables, display ads, logos, signatures, icon buttons, borders, lines, and custom fonts. Because of the Web's ability to display graphic images with the use of a graphical browser, you can design a unique, original Web page that communicates your message effectively. First let's examine the various graphic image file formats.

Graphic Image File Formats

Graphics used for Web-based artwork are called **bitmapped** (or **raster-based**) graphics, in contrast to object-oriented (or vector-based) graphics. Bitmapped graphics form images with a series of **pixels**, a series of interconnected dots. Graphic images that are produced by image processing, paint, scanning, and video capture applications are bitmapped. File formats that use bitmapped graphics include: Tagged Image File format (TIFF), Z-soft Graphic Image format (PCX), Windows Bit-mapped format (BMP), Graphical Interchange format (GIF), and Joint Photographic Experts Group format (JPEG).

Web browsers will accept two types of graphic image files: GIF and JPEG files.

GIF. The **graphical interchange format** (**GIF**) was originally developed for CompuServe to facilitate the transfer of graphics among the various computer brands connected to the service (i.e., Apple II, Atari, Commodore, IBM Macintosh, Amiga, UNIX, and X-Windows).

GIF is an 8-bit, bitmapped, 256-color format and works well with graphic lines or separator bars; icons, buttons, or bullets for lists; charts, tables, or graphics you want to place with text; small illustrations; or thumbnails. UNISYS presently holds a patent on the compression routine that is a part of the GIF format. Developers of programs that use a GIF format do not currently need to pay a royalty, but it is unclear whether this will change in the future. If royalties are required, a new format may need to be developed to avoid the legal problems associated with using GIF files. An animated GIF format exists (see Chapter 6).

JPEG. The **joint photographic experts group** format (**JPEG**) is an 8- or 24-bit graphic image file format that takes advantage of human perceptual characteristics and does not deal with less essential information. JPEG works best with complex images that use more than 256 colors; images you don't need to include directly on the Web page; large images; and photograph-quality images. Because JPEG uses a more complex compression scheme, its file sizes are often much smaller than GIF files.

Using Graphic Images

The graphic images included in a Web page can be one of two types: inline or external. **Inline images** are those that can be displayed directly on a Web page; **external images** must be downloaded separately and are usually accessed through a link. Inline images can be displayed automatically by most graphical browsers. External images, however, require a helper application program to be configured with your browser in order to be viewed.

INLINE IMAGE ADVANTAGES

Inline images give you some control over the format of the presentation, and viewing them is guaranteed since all inline images use either GIF or JPEG formats. In addition, inline images can be saved via a special Save Image to Disk command in some browsers such as Netscape. Finally, the browser will automatically load each inline image so that you do not need to wait for an external helper application to launch and decode the image.

EXTERNAL IMAGE ADVANTAGES

Because external images must appear in separate windows, users have more control over the size, location, and other features of the image window. Moreover, external images do not limit you to GIF or JPEG formats; they can use any format supported by your helper applications. Files can be saved for later manipulation and viewing, allowing you to access higher-resolution images.

IBM AND IBM PC-COMPATIBLE HELPER APPLICATION VIEWERS FOR EXTERNAL IMAGES

As we saw in Chapter 2, you may need to install helper applications for your browser to display graphic images. Do you remember why helper applications are needed? When Netscape or another browser retrieves a file with a format it cannot read, the application attempts to use an external helper application capable of reading the file. Several helper application viewers for the IBM and IBM PC–compatible computer can be used for external images.

LView. One of the most popular is LView and LView Pro, by Leonardo Loureiro. You need a 386 or better processor to run it, and a Super VGA card and monitor are recommended; a 32-bit Windows or Windows NT version requires a 486 processor. LView is able to read or write the following graphic formats: GIF, GIF87a, GIF-89a, JPEG, JPG, JFIF, TGA, BMP Windows and OS/2, PCX, PPM, PGM, TIF, and TIFF.

Once you have downloaded a graphic image with LView, you can manipulate it by using the following features: resize, redimension, crop, add text, flip horizontal, flip vertical, rotate left, and rotate right. LView gives you the ability to save the downloaded file to a different graphic file format. The freeware version of LView can be downloaded at ftp://ftp.igc.net/pub/igc/viewers/view31.zip. The shareware version (LView Pro) can be downloaded at ftp://ftp.igc.net/pub/igc/viewers/view1b.zip. Figure 4-1 shows the LView freeware viewer.

WinGIF. WinGIF is another easy-to-use helper application for viewing external images. It can read the following graphic file formats: GIF, GIF87a, GIF89a, BMP Windows, RLE, and PCX. WinGIF is capable of manipulating downloaded graphics by using the following features: resize, mirror, flip, and rotate. WinGIF can be downloaded at ftp://gatekeeper.dec.com/pub/micro/msdos/win3/desktop/. Figure 4-2 shows the WinGIF viewer.

Figure 4-1. The LView Freeware Viewer

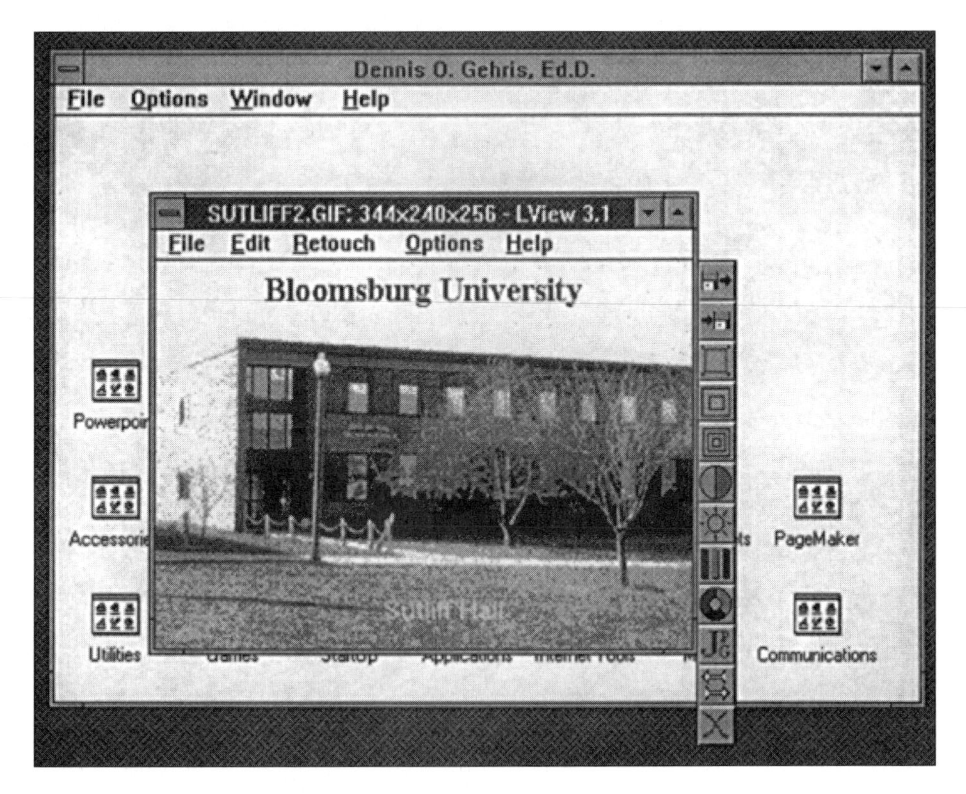

Figure 4-2.
The WinGIF Viewer

MACINTOSH HELPER APPLICATION VIEWERS FOR EXTERNAL IMAGES

If you are using a Macintosh computer, various helper application viewers are available. The two recommended viewers are JPEGView and GIFConverter.

JPEGView. JPEGView, by Aaron Giles, requires a System 7.0 or higher operating system and Quicktime and Color QuickDraw programs. The viewer supports JPEG, PICT, GIF, TIFF, BMP, MacPaint, and Startup Screen file formats. It can be downloaded at http://www.gsn.org/web/macware/jpegview.html.

GIFConverter. GIFConverter, by Kevin Mitchell, requires System 6.0.5 or higher operating system, but Quicktime or Color Quickdraw is not required. The program is shareware and a $40 registration fee is requested. The viewer supports the following file formats: GIF, JPEG, TIFF, RIFF, PICT, MacPaint, and Thunderscan. It can be downloaded at http://www.macworld.com/cgi-bin/software.pl/Graphics/software.32.html.

CONFIGURING NETSCAPE

As an example, here are the steps necessary to configure Netscape for the LView viewer on an IBM or IBM PC–compatible computer. If you don't configure Netscape for a helper application viewer, when Netscape encounters a file format for which a helper application viewer has not been configured you will be prompted to configure a viewer at that time.

1. Download the viewer and unzip the files, if necessary.
2. Place the files in a subdirectory called C:\LVIEW (use another drive letter, if you wish).
3. Start Netscape.
4. Click on General on the Options menu.
5. Click on the Helper Applications folder.
6. Click on the scroll bar under File type until you see the file type image/tiff. Click to select it.
7. Click on Launch the Application under Action.
8. Click an insertion point on the white typing area to the left of the browse button and key C:\LVIEW\LVIEW31.EXE (or other drive letter). Click on OK.
9. Repeat steps 4, 5, and 6.
10. Compare your screen with Figure 4-3.

You can add additional graphic image file types by clicking on the Create New Type button. For example, key IMAGE for the Mime Type and PCX for the Mime SubType to read PCX files by following the steps just listed.

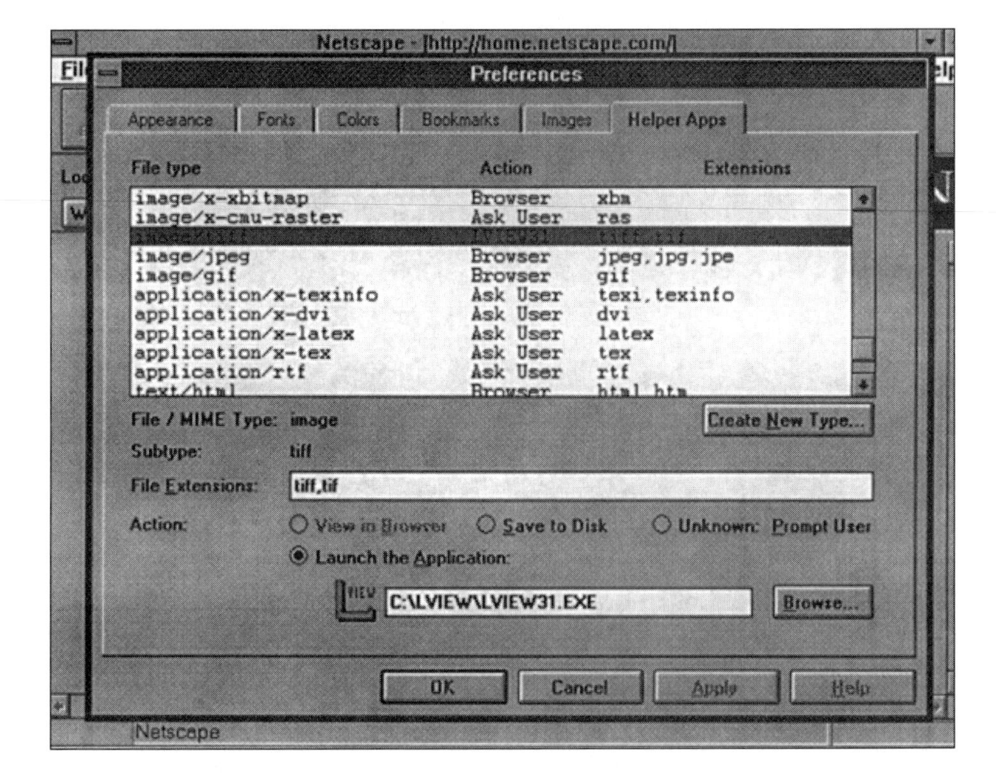

Figure 4-3. LView Helper Application Installed

Obtaining Graphic Image Files

Now that you know how to configure Netscape to display graphic images, you may need to be able to locate the graphic files. There are several methods to obtain graphic image files that you can save on a disk to use later on a Web homepage that you're developing or use in some other word processed document. These methods include using computer clip art, finding an Internet site with a graphic image that you like and downloading the graphic in a file, creating graphic image files with a scanner, and creating graphic image files using video capture. We will examine each of these methods and provide steps for using them.

A note of caution before we begin: You should be careful about using a graphic image that may be copyrighted or has other legal restrictions placed on it. Many authors and artists post a statement indicating that the document is copyrighted and give permission for distributing it in an electronic form as long as it isn't sold or made part of a commercial venture. Laws protect the copyrighted works of others, even if a similar statement is not provided. It is always a good idea to give the artist credit before you display a copyrighted image and check to see if there is any mention of a fee required to use the image.

COMPUTER CLIP ART

You've probably heard about or have even used clip art, a library of graphic files stored on diskette or CD-ROMs that can be used in word processing, desktop publishing, or other application programs. Most computer software stores sell clip-art packages that include a license to reproduce the images. Clip art is saved in various file formats and its quality varies from very good to mediocre to poor. Today many clip-art collections sold on CD-ROM discs contain an amazing quantity of image files that are often poor quality. The file format must be one that is supported by the application program you are using. If you want to use the files on a Web page, remember that you are limited to GIF and JPEG, although it may be possible to use a utility program to convert the files to the desired format. Some drawing application programs such as CorelDRAW include huge clip-art libraries.

Finding and Downloading Graphic Images on the Internet

Another method of obtaining graphic image files is to find and download them from the Internet. The Internet contains many sites that serve as depositories of graphic files of many types and covering many subjects. Individuals and organizations provide the graphic files free to anyone who wants them. The best approach is to locate the sites that allow you to do a search on a filename or subject. Otherwise, you'd have a difficult time in locating the files, especially since the filenames do not describe the nature of the image very well. After viewing the files on the browser preview screen and deciding that you want to download a particular graphic, Netscape and other browsers allow you to save the graphic file onto your hard drive or floppy disk. Table 4-1 provides a list of sample URLs for sites that allow you to find graphic image files. It should be noted that many more sites on the Internet contain graphic image files. You might want to use a search engine or resource tree site to locate additional URLs.

TABLE 4-1 SAMPLE INTERNET SITES FOR GRAPHIC IMAGE FILES

Site Name	URL
WWW Images Examples	http://www.cit.gu.edu.au/images/
American Memory	http://rs6.loc.gov/amhome.html
Sunet Image Archive	ftp://ftp.sunet.se/pub/pictures
Isca Image Archive	ftp://grind.isca.uiowa.edu/image

HTTP URL SITE

As an example of using an http URL for finding a graphic image file, we will use the http://wuecon.wustl.edu/other_www/wuarchimage.html site to show the steps for downloading a graphic image file of a Dalmatian. The procedure will differ if you use another site.

1. Start Netscape.
2. Click on the Open button and key http://wuecon.wustl.edu/other_www/wuarchimage.html. (Note: an alternate URL is http://www.cm.cf.ac.uk/Misc/wustl.html.)
3. After the main page is loaded, scroll down to the Description typing box and click an insertion point in it. Key dog.
4. Scroll down to the Submit Query button and click on it.
5. After the next page is loaded, click on the Submit This Query button.
6. Click on lulu. (You can click on another filename if this one is not available.) After downloading, the file should be displayed as shown in Figure 4-4.
7. Click on Save As... on the File menu. Choose a path for the lulu.gif file and click on OK. (A saving location box should appear and the file will be saved.)

Figure 4-4. Lulu.gif File Displayed After Downloading

8. To check to see if the file has been saved and is readable, click on Open File... on the File menu. Click on All Files [*.*] under List Files of Type. Choose the path that you wish to use to save the file. Click on the lulu.gif filename and click on OK. The file should be displayed on the preview screen. (Note: You could also read this graphic image file with one of the helper application viewers.)

FTP URL SITE

As an example of using a ftp URL for finding a graphic image file, we will use the ftp://grind.isca.uiowa.edu/image site to show the steps for downloading a graphic image file of a clown. The procedure will differ if you use another site.

1. Start Netscape.
2. Click on the Open button and key ftp://grind.isca.uiowa.edu/image.
3. After the page is loaded, click on /gif.
4. After the page is loaded, click on /image.
5. After the page is loaded. click on clown.gif (You can click on another filename if this one is not available.) After downloading, the file should be displayed as shown in Figure 4-5.

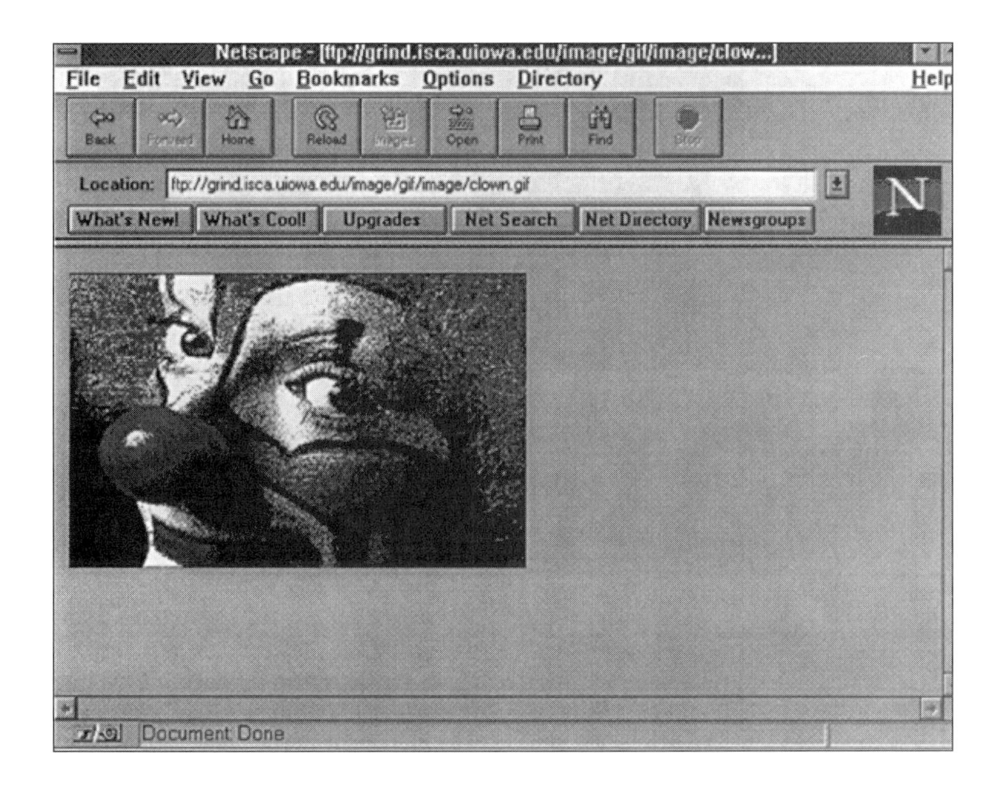

Figure 4-5. Clown.gif File Displayed After Downloading

6. Click on Save As... on the File menu. Choose a path for the clown.gif file and click on OK. (A saving location box should appear and the file will be saved.)
7. To check to see if the file has been saved and is readable, click on Open File... on the File menu. Click on All Files [*.*] under List Files of Type. Choose the path that you wish to use to save the file. Click on the clown.gif filename and click on OK. The file should be displayed on the preview screen. (Note: You could also read this graphic image file with one of the helper application viewers.)

MOUSE BUTTON METHOD

If you have an IBM or IBM PC–compatible computer and Netscape, you can use the right mouse button for saving a graphic image. On the Macintosh computer, hold down the mouse button for about one second. Choose Save This Image As... from the popup menu. This is a really nice way to download the file of an image that you happen to see on a Web page. However, remember the previous discussion about legal issues in downloading a file without permission. As an example, let's assume that we want a graphic image file of an eagle. We use a search engine and find the URL http://www.soar.com/~info/Eagles/Eagles.htp. The following steps are used to save a graphic file of an eagle.

1. Start Netscape.
2. Click on the Open button and key http://www.soar.com/~info/Eagles/Eagles.htp.
3. After the page is loaded, move the mouse arrow on the first picture and click the right mouse button (or hold the button on the Macintosh). Compare your screen with Figure 4-6.
4. Click on Save This Image As... on the popup menu that appears. Choose a path for the facingus.jpg file and click on OK. (A Saving Location box should appear and the file will be saved.)
5. To check to see if the file has been saved and is readable, click on Open File... on the File menu. Click on All Files [*.*] under List Files of Type. Choose the path that you wish to use to save the file. Click on the facingus.jpg filename and click on OK. The file should be displayed on the preview screen. (Note: You could also read this graphic image file with one of the helper application viewers.)

If you are using Netscape, it is also possible to save the location of a graphic file to the Windows clipboard. This makes it possible for you to have a browser read the graphic file from the remote location. Click the right button on a graphic (or hold the button with a Macintosh) and choose the option Copy This Image Location on the popup menu. You can then paste the copied location

Figure 4-6. Clicking the Right Mouse Button

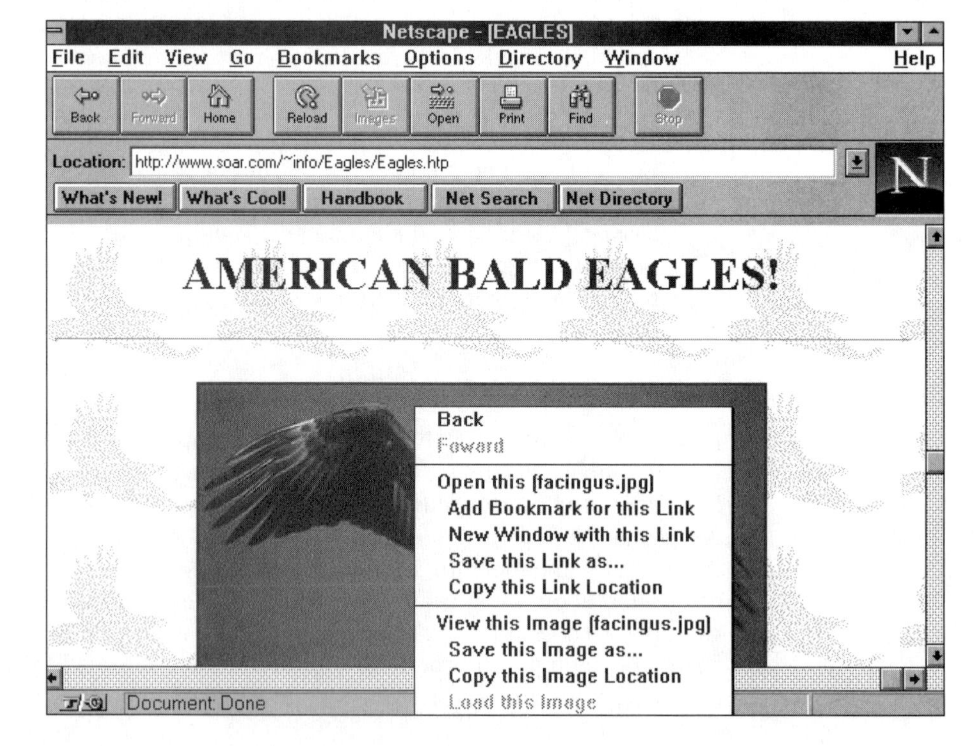

from within your favorite HTML editor. First place the <IMG SRC =" "> tag on the screen, click an insertion point between the two quotation marks, and choose Paste from the Edit menu. The location path should properly appear in the inline graphic tag. You can adjust where the image will appear on the browser preview area by specifying the alignment in the tag (bottom, middle, top) or by enclosing the entire tag in <CENTER></CENTER> tags to center the graphic image.

Next, we will explain how to use a scanner to create graphic image files. You can use this method if you have access to a scanner that is connected to a computer on which an image software package with a scanner module has been installed.

Creating Graphic Image Files with a Scanner

Another method of obtaining graphic image files is to use a hardware device known as a **scanner** to convert digitized images into GIF or JPEG file format for use in a Web page. You can also convert images to one of the other popular file formats (TIFF, PCX, etc.) for use in a word processing or desktop publishing program. Scanners are available from various manufacturers, including Microtek, Hewlett Packard, Okidata, and Epson. There are three types of scanners: hand-held, sheet-fed, or flat-bed. The hand-held scanner is dragged across the document. It is the least expensive, but you are restricted by the size of the document that can be scanned. Sheets of paper containing the drawing or photograph to be scanned are fed into the sheet-fed scanner. The most popular scanner is the flat-bed scanner, which looks very similar to most photocopiers. You lay the document to be scanned face down on the glass bed of the scanner before performing the scan.

A good scanner allows you to input various types of images: drawings, photographs, magazine pages, or anything else that is flat. It is also possible to use scanners to scan text and save it to a file. The scanner works in conjunction with scanner software, which often comes packaged with the scanner when it is purchased.

Various image software packages contain scanning modules, including Paint Shop Pro, Adobe Photoshop, Hewlett Packard DeskScan, and Aldus PhotoStyler. These programs are *raster format image editing programs*, a type of software more commonly known as a *bitmap editor*. Raster image formats break a picture into a grid of equally sized pieces, called *pixels*, and record color information for each pixel. Common examples of raster formats include the Windows .BMP format and the CompuServe .GIF format. Paint Shop Pro, for example, isn't confined to a short list of raster file formats. It provides full support for all of the most popular raster formats, full or partial support for many less-popular raster formats, and can read nine meta and vector image formats.

Once the scanner driver and accompanying software have been installed for the scanner you are using, it may be possible to scan through a word processing or desktop publishing program that you are using. For example, WordPerfect 6.1 for Windows, a popular word processing program, contains the WordPerfect Draw module that can be used to scan an image. With the addition of the Internet Publisher add-on, you can create a HTML document in WordPerfect (see Chapter 3).

Adobe PageMaker 6.0, a well-known desktop publishing program, supports TWAIN, a cross-platform interface that lets you create a TIFF image using a device (such as a scanner, video-capture board, or digital camera) attached to your computer and import the image into your publication without leaving PageMaker. Your device must support TWAIN for you to take advantage of this feature. PageMaker 6.0 has a HTML plug-in module that allows you to convert PageMaker documents into HTML format.

Adobe Photoshop is an excellent and popular program not only for scanning but for graphic file manipulation. It is available for both IBM or IBM PC–compatible and Macintosh computers with many advanced features. The following steps relate to using Adobe Photoshop to scan a drawing and save the resulting file in GIF format. To follow all these steps, you must have a computer with Adobe Photoshop software that is attached to a flat-bed scanner.

1. Turn on the scanner.
2. Select a page containing line art drawing and place it face down on the bed of the scanner. The top of the page should face the front of the scanner.
3. Find and load the Adobe Photoshop program. It should be in a special group set up for the program.
4. After the program is loaded, select Acquire Image from the File menu. Click on Twain in the box (see Figure 4-7). (Note: It is assumed that the scanner has been configured for the software. If it hasn't been properly installed, you may get an error message.) The scanner screen should appear, and it may show the last image that was scanned in the scan window.
5. At the top of the screen, select Line Art as the scanning mode. (Note: If you are scanning a color photograph, choose the Photograph mode; if you are scanning a black and white photograph, choose either the Grayscale or Halftone modes.)
6. Click on the Prescan button. This should activate the scanner, and the program may show the amount yet to be scanned with a slider indicator. After the scanning process is completed, the document should appear in the scan window. (Note: A different image will appear in the scan window.)
7. Click on the Image Selector tool on the toolbox if necessary and carefully position a selector box around the portion of the document that contains the graphic image that you want to save to a file (see Figure 4-8).

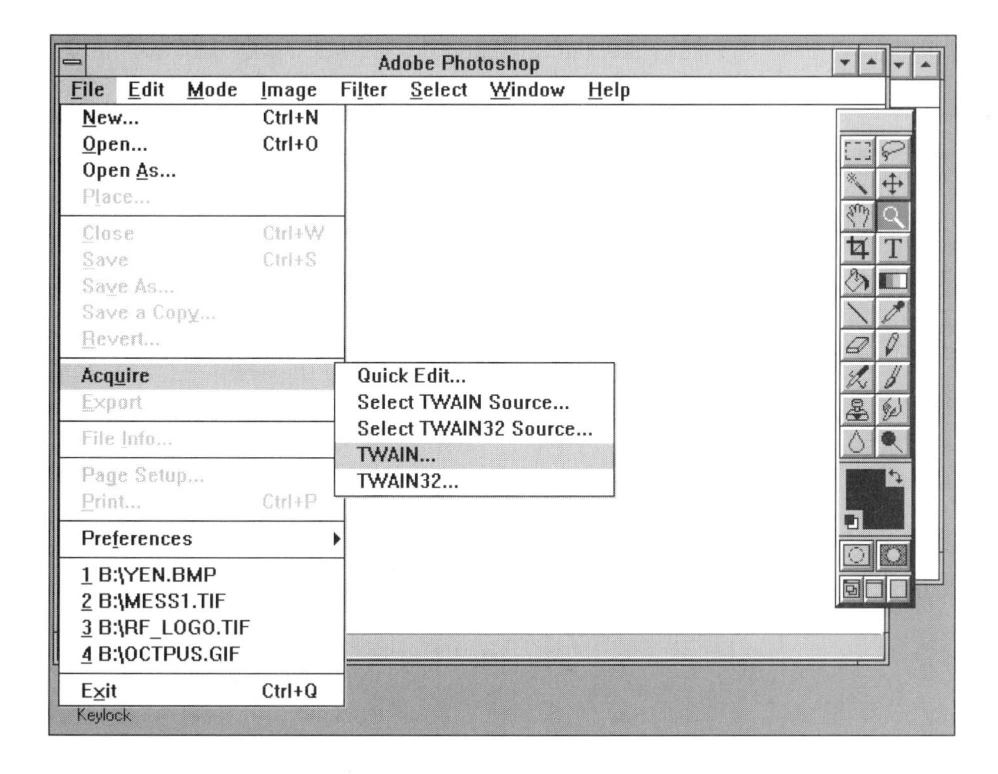

Figure 4-7. The Adobe Photoshop Main Screen with File Menu

Figure 4-8. Adobe Photoshop with Selector Box Positioned

8. Click on the Scan button. This should activate the scanner, and the program may show the amount yet to be scanned with a slider indicator. After the scanning process is completed, click on OK and a box should appear containing the image.
9. Click on Save As on the File menu and a Save box should appear. Select GIF from the file type selector. Note the other file formats that are available. Select a path and key a filename and click on OK. If the file was saved, the filename should appear at the top of the image box (see Figure 4-9).
10. Exit the program and turn off the scanner.

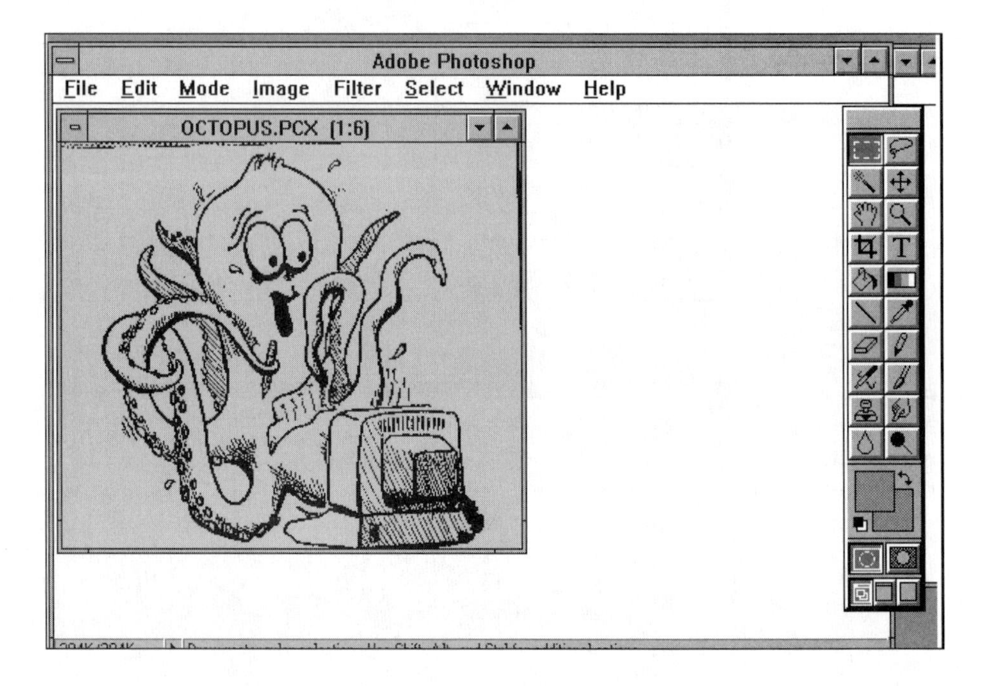

Figure 4-9. Adobe Photoshop with Image Box

Creating Graphic Image Files via Video Capture

A final method of acquiring graphic image files is **video capture**. This technique requires you to purchase several pieces of peripheral equipment and specialized software if you are using an IBM or IBM PC–compatible computer. You need a **video capture card** or device installed in one of the expansion slots or plugged into a port in the computer, a video camera, and software that enables you to capture still images from the video source. Several manufacturers produce video capture cards. One is the Captivator card by the VideoLogic Company, which captures both still images with their VideoSnap software and moving images in conjunction with Microsoft Video for Windows. Both pieces of software come packaged with the VideoLogic card. Other manufacturers offer packaged software with their video capture cards.

The alternative to a video capture card is a device that attaches to the printer port at the back of your computer. Snappy, distributed by Minolta USA, is a popular device that offers high-resolution video capture (1500 x 1125). Images can be captured in standard file formats in literally millions of colors.

You can attach any video camera with an output cable attachment to the video capture card. If you will also be using the camera for videoconferencing, you may want to purchase a desktop camera. One example of a desktop camera is the FlexCam by Videolabs, whose lens is mounted on a flexible gooseneck stem. It also contains microphones that can be used to record sound along with your moving video images. Other desktop cameras are placed on the top of the monitor or mounted on a stand next to the computer.

As an example of how video capture software is used to capture still graphic images, we'll examine the procedures for using VideoSnap software, which are similar to the steps for using other software of this type. When you turn on the attached video camera and load the software, a live video window appears. You can then capture the window as a still image by clicking on the Image Snap button, the first button at the left on the toolbar. The Live Video window becomes Image 1 box, which can then be saved by choosing Save As on the File menu. You can use this file on a Web page or in a word processing or desktop publishing program. VideoSnap will save images in the BMP file format only. You can convert the file to GIF or JPEG by using a utility program like Lview. After you capture the image, another live video window appears that shows the current image from the camera (see Figure 4-10).

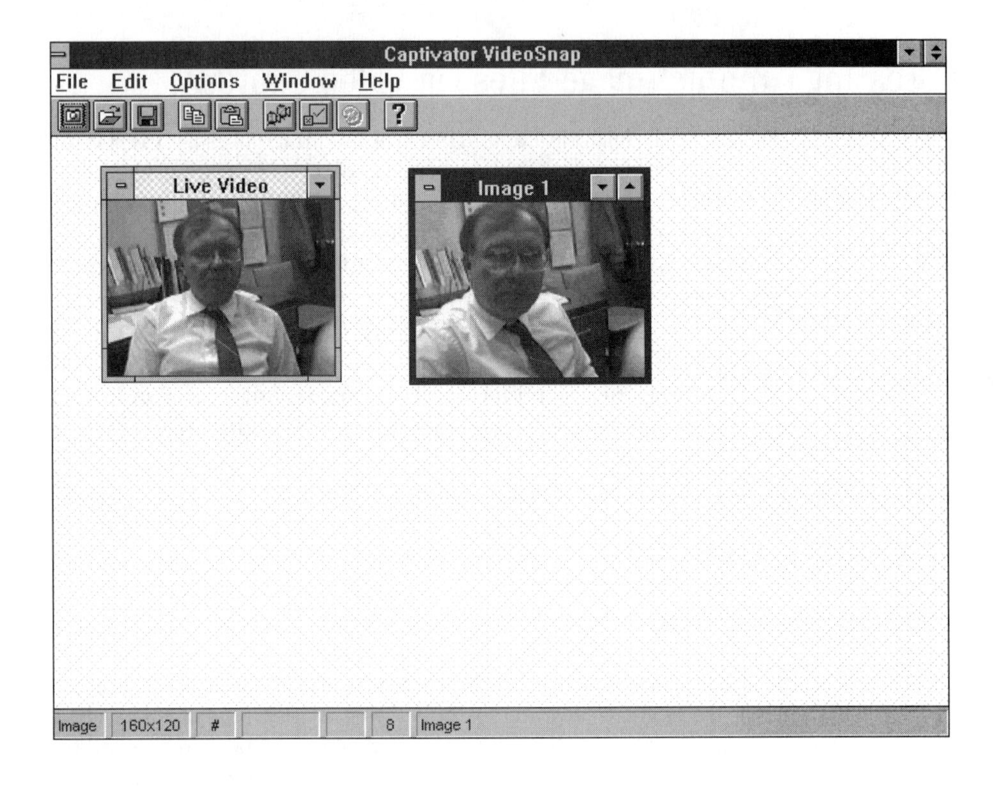

Figure 4-10. The VideoSnap Program

Questions for Review

1. A graphic image can consist of what types of elements?
2. What are bitmapped graphics?
3. What two types of graphic image file types will Web browsers accept?
4. For what types of graphic images are GIF image files used?
5. For what types of graphic images are JPEG image files used?
6. What is an inline image?
7. What are the advantages of using inline images in Web documents?
8. What is an external image?
9. What are the advantages of using external images in Web documents?
10. What are the functions of helper application image viewers? Name the two most popular helper application viewers for the IBM or IBM PC–compatible computer and two for the Macintosh computer.
11. How do you configure Netscape for a helper application image viewer?

12. What methods can be used to obtain graphic image files?
13. What is computer clip art?
14. How are graphic images on the Internet found and downloaded?
15. What are the advantages of using raster format image editing programs? Name two examples of this type of program.
16. How can a scanner be used to create a graphic image file?
17. How can graphic image files be created with video capture?

Exercises

1. Use Netscape or another browser to download one of the application helper application image viewers using the URLs provided below. Unzip the file, if necessary, and configure Netscape or another browser to read TIFF and PCX file types.

   IBM and IBM PC–Compatible

LView	ftp://ftp.igc.net/pub/igc/viewers/lview31.zip
LView Pro	ftp://ftp.igc.net/pub/igc/viewers/lviewp1b.zip
WinGIF	ftp://gatekeeper.dec.com/pub/micro/msdos/win3/desktop/

   Macintosh

JPEGView	http://www.gsn.org/web/macware/jpegview.htm
GIFConverter	http://www.macworld.com/cgi-bin/software.pl/Graphics/Software.32.html

2. Use one of the Internet sites shown in Table 4-1 to find and download graphic image files of a horse, computer, palm tree, or any other image of your choice.

3. If you are using an IBM or IBM–PC compatible computer, use the mouse button method to download at least one graphic image file using any of the following URLs.

Icons for Web Pages	http://www-ns.rutgers.edu/doc-images/icons
Grand Canyon National Park Photos	http://www.kbt.com/gc
Canyonlands National Park Area	http://pizero.colorado.edu/~ewv/canyonlands.html
Pennsylvania Dutch Country	http://www.800padutch.com

4. If you are using an IBM or IBM PC–compatible computer, use the mouse button method to save the location of at least one graphic in one or more of the locations in item 3. Use an HTML editor program to paste the location path into a inline image tag. Test the HTML file in a browser to see if the image appears.
5. If you have access to a scanner, select a drawing or photograph and use the scanner to create a file in the GIF or JPEG format. If you would like to use it in a word processing or desktop publishing program, save the file in the PCX, TIFF, or BMP format.
6. If you have access to a computer with video capture capabilities, use a video camera to capture a photograph of yourself or a friend and save it in the GIF format.

CHAPTER

5

Audio on the Internet

In this chapter we will cover the use of sound on the World Wide Web to enhance communication and understanding. The chapter will describe configuring a PC for sound, audio file formats, Netscape LiveAudio, and audio helper applications, and it will give step-by-step instructions for downloading and using the RealAudio and TrueSpeech real-time audio systems.

What You Will Learn

- How audio is used
- Types of audio file formats
- How to configure a PC for sound
- How to use Netscape LiveAudio
- Types of audio helper applications
- How to download and use the RealAudio Player
- How to download and use the RealAudio Encoder
- How to download and use the TrueSpeech Player

Using Audio

As we saw in Chapter 1, multimedia is a combination of text, graphics, audio, and full-motion video. **Audio** or sound, the third element, can consist of human speech, music, or sound effects. Human speech on a Web page helps to personalize the text and graphics. Music can provide a means for understanding and appreciating cultural differences. Music and sound effects together give the Web page novelty and entertainment value, which helps to stimulate interest.

There are some practical reasons for providing sound on a Web page. One is that you may want to include resources that are intrinsically sound based, such as speeches, music, news reports, or sound art. In addition, text can be annotated with a spoken commentary. You may also want to include sound effects relating to a natural phenomenon, such as a tornado, hurricane, or volcano, or annotate a discussion of wild birds with bird calls. Coverage of historical events is enhanced by a spoken reenactment or news report.

Audio File Formats

There are five main audio file formats: Macintosh Audio Information File format (AIFF), Audio-file format on UNIX machines (SND, AU), Microsoft Windows Audio format (WAV), and Musical Instrument Digital Interface format (MIDI or MID). The Macintosh computer uses the AIFF file format; UNIX-based systems use SND and AU files. PC-compatible computers running Windows use WAV, and all platforms can use MIDI files. Although MIDI is the standard for professional electronic music, these files are not used on Web pages for several reasons. First, MIDI files are not suitable for nonmusical sounds such as speech. Second, helper applications capable of playing MIDI files are not widely available. Many Web sites include sound files in several formats; you are given a choice of which file format you would like to select. Table 5-1 lists the audio file formats.

Many servers on the Internet contain sound files. Table 5-2 provides a list of some sample Internet sites for sound files.

TABLE 5-1 THE AUDIO FILE FORMATS

Format	Description
AIFF	Macintosh Audio Information File format
SND, AU	Audio-file format on UNIX machines
WAV	Microsoft Windows WAVE audio
MIDI	Musical Instrument Digital Interface

TABLE 5-2 SAMPLE INTERNET SITES FOR SOUND FILES

Site Name	URL
Classical Midi Archives	http://www.prs.net/midi.html
Sound Clips Page	http://www.amscomm.com/monsters/sound.html
SunSITE Sound Files	ftp://sunsite.unc.edu/pub/micro/pc-stuff/sounds
Josh's Sound Files	http://www.midcoast.com/~packard/sfile.html
Bitboard Show Prep Networks	http://www.bitboard.com/sound.htm

Configuring a PC for Sound

As we saw in Chapter 2, a sound card is usually a requirement for using sound on a PC-compatible computer. When the sound card is inserted into one of the expansion slots, accompanying software containing drivers permits the computer to use the sound card. External speakers are attached to the sound card, and many cards possess a jack to which a microphone can be attached to permit sound recording. If you are using Windows 3.1, there is a way to play sounds you encounter on the Internet without a sound card. The speak.exe file allows you to configure your computer so that sound files will be played through your PC's internal speaker. Although it will not produce quality sound, this is a good alternative for those PCs not equipped with a sound card. The internal speakers in some multimedia PCs can be upgraded to speakers that will provide a richer sound.

The speak.exe file can be downloaded at one of the following URLs: ftp://ftp.microsoft.com/softlib/Mslfiles or ftp://hubcap.clemson.edu/pub/pc_shareware/windows/Drivers/speak.exe. You may also use the Microsoft Download Service (MsDL) at (206) 936-6735. Place the file in a temporary subdirectory of your hard disk or on a floppy disk. Run the self-executing speak.exe file. Several files will result, including speaker.drv, which is the driver file. Click to select the Main group, Control Panel, Drivers, and click on the Add button. When you are requested to insert the disk containing the setup information, insert the disk you made during the downloading procedure or type the path of the directory where you downloaded the driver. A dialog box will appear the first time you install the driver, allowing you to adjust the options for the driver. You can adjust the speed, volume, and number of seconds to limit the playback. It is a good idea to choose No Limit so that an audio file of any length can be played. Also, you usually need to check the Enable Interrupts box before you click on OK (see Figure 5-1).

If you are using Windows 95, it is also possible to use your computer's internal speaker. If no sounds are playing through the internal speaker, you will need to locate the speak.exe file using one of the sources previously provided. In My Computer or Windows Explorer, double-click the file you downloaded to extract the driver in a temporary subdirectory on your hard disk. Click the Start button, point to Settings, and then click Control Panel. Double-click the Add New Hardware icon, and then follow the onscreen instructions. At the prompt "Do you want Windows to search for your new hardware?" click No. Select Sound, Video, and Game Controllers, click Have Disk, and then specify the location of the driver file. Click on Sound Driver for PC Speaker, OK, and Finish. Restart your computer, when you are prompted to do so.

In the Control Panel, double-click on the Multimedia icon. On the Advanced tab, double-click on the Audio Devices branch, double-click on Audio for Sound Driver for PC Speaker, and click on Settings. Follow the instructions previously provided for adjusting the settings. Click on OK until you return to the Control Panel.

The PC speaker driver plays only WAV files. The current version of Media Player included with Windows 95 does not play WAV files with the PC speaker driver. You must use the Sound Recorder to play WAV files. The built-in Windows 95 Volume and Mixer tools will not control the PC speaker driver.

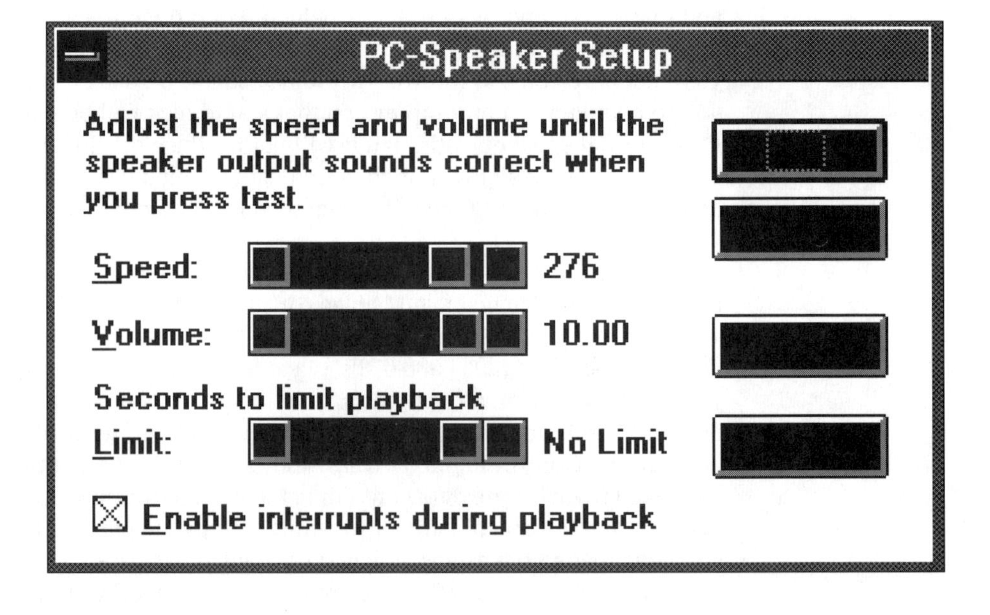

Figure 5-1. The PC Speaker Setup Dialog Box

Netscape LiveAudio

Netscape 3.0 contains the LiveAudio feature, which includes native support for the standard sound formats AIFF, AU, MIDI, and WAV. With this feature you are able to play and hear sound files embedded in HTML documents, making it unnecessary to download a sound helper application to play sound files. By clicking on a sound file, LiveAudio instantly provides an easy-to-use player console. Simple visual controls for play, pause, stop, and volume are included. Figure 5-2 shows the LiveAudio player console.

Because you may be using an earlier version of Netscape or some other browser that needs audio helper applications, we'll examine some of the popular audio helper applications for both the IBM or IBM PC–compatible and Macintosh computers.

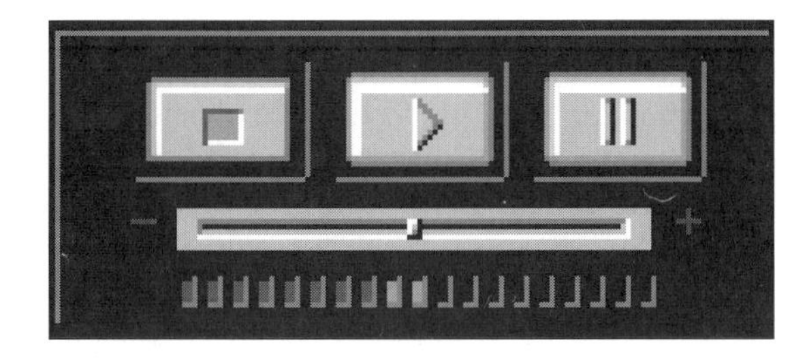

Figure 5-2. The Netscape LiveAudio Player Console

Audio Helper Applications

To hear sounds on the Web using both IBM or IBM PC–compatible or Macintosh computers and a browser like Netscape 2.0, you may need an audio helper application program. Fortunately, various helper applications are available on the Internet FTP servers. These sound application programs execute when an audio file that is supported is downloaded from the Web. In addition, they allow you to control playback, save sounds to a local hard disk, and convert various sound formats. These sound application helper programs differ in the number and types of audio formats they support. The following are the most frequently used programs for both the IBM or IBM PC-compatible and Macintosh computers.

IBM AND IBM PC-COMPATIBLE COMPUTERS

NAPLAYER. The Netscape Audio Player (NAPLAYER) is installed automatically when you install the PC-compatible version of Netscape (versions 1.1 to 2.0). NAPLAYER plays sound files in the AU and AIFF formats. You can obtain additional information on this player at http://home.netscape.com/newsref/win audio.htm. Figure 5-3 shows the Netscape Audio Player when it is executed by clicking on the NAPLAYER icon in the Netscape Group Window. The same interface is displayed within Netscape when a sound file is downloaded from the Internet. To play a sound, click on the Play (right arrow) button.

WHAM. The Wave Hold and Multiply Audio Player (WHAM) plays files in WAV, VOC, IFF, AIFF, AU, and raw formats. As with NAPLAYER, you play a sound by clicking on the Play (right arrow) button. WHAM will also convert to and record sounds in these formats. To convert a sound file to another format, simply load the original sound file, choose Save As from the file menu, choose the new type of file, key in a new filename with appropriate extension and path, and click on OK. To record a sound file, you must have an installed sound card with an attached microphone. To record a sound, choose Record New from the file menu and click on Record. After memory space has been allocated, record your sound, then click on Stop and OK. You can play back your recorded sound by clicking on the left arrow on the bar at the top of the screen. To save your sound, choose Save As on the File menu. WHAM is available for downloading at ftp://sunsite.unc.edu/pub/micro/pc-stuff/sounds/.cap/wham.zip or some other site. Figure 5-4 shows the WHAM player with a loaded WAV file.

WPLANY. The Will Play Any Audio Player (WPLANY) is another audio helper application available for free download. It will detect and play any sound through a Windows 3.1 audio device in the following file formats: VOC, AU, WAV, SND, and IFF. WPLANY does not have a player bar that NAPLAYER or WHAM possesses. Sounds will automatically play without needing to click on a Play button. There is also no capability for sound conversion or recording. It is available for download at ftp://uxa.ecn.bgu.edu/pub/micro/ibmpc/win/helpers/wplny/wplny11.zip or some other site.

Figure 5-3. The Netscape Audio Player

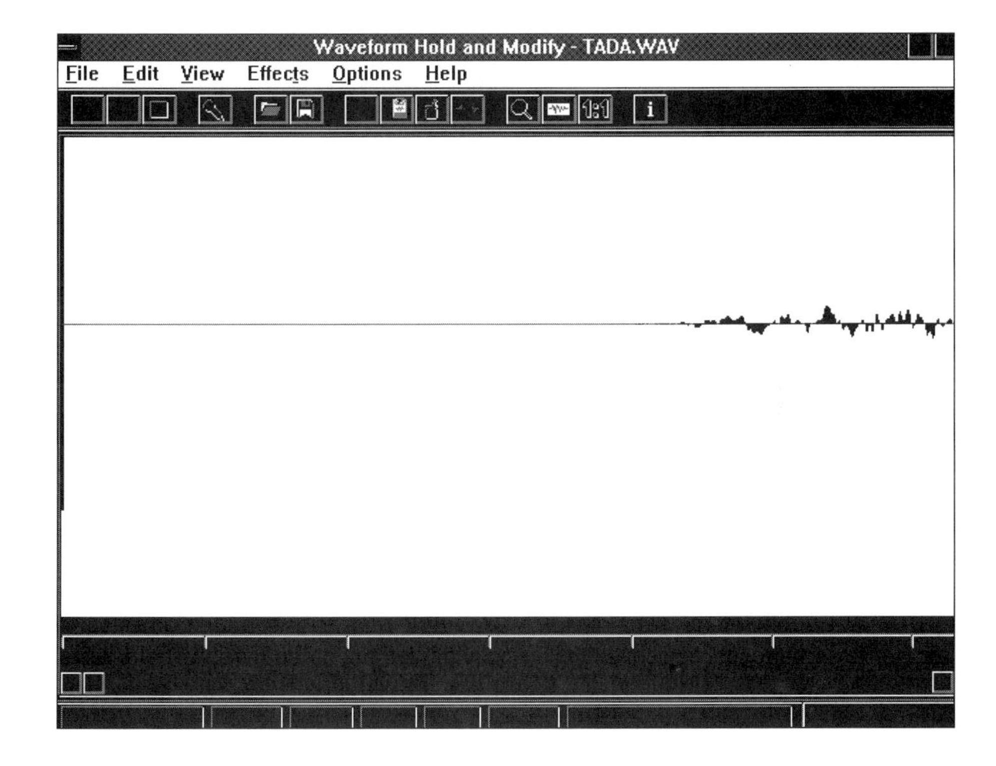

Figure 5-4. The WHAM Player

Do you use a Macintosh computer? The following is a description of several audio helper applications for the Macintosh. Keep in mind that if you are using Netscape 3.0 or above, you may not need to use one of these.

MACINTOSH COMPUTERS

The two most commonly used audio helper applications for the Macintosh are SoundMachine and SoundApp. SoundMachine opens, plays, and records files in AIFF, System 7 sound, and AU file formats. It is available for downloading at ftp:wcarchive.cdrom.com/.12/mac/umich/powermac or some other site. SoundApp opens, plays, and can convert sound files in AIFF, Soundedit, System 7 sound, Quicktime sound tracks, and UNIX AU (all variants), and Windows WAV. It is available for downloading at ftp://ftp.utexas.edu/pub/mac/sound/ or some other site.

Real-Time Audio Players

Did you know that you may be able to listen to the current news, weather, or sporting event by using your Web browser? A recent development in the use of sound on the Web is the inclusion of what can be termed **real-time audio players**. These helper applications let you listen to a live and rebroadcast audio on the Internet in real time without needing to download, store, and play a sound file. This is much like turning a Web connection into a radio in which the sound and music quality is equivalent to hearing an FM (mono) transmission.

With a real-time audio player you can listen to breaking news stories, sporting and entertainment events, corporate meetings, and the like. Two of these players are currently available for free download: RealAudio from Progressive Networks, Inc. and TrueSpeech from DSP Group, Inc. The URL for RealAudio is http://www.realaudio.com and the URL for TrueSpeech is http://www.dspg.com/internet.htm.

RealAudio 2.0 or 3.0 requires a 486/66 or faster computer with Windows 95/NT or a computer with a 486/66 or Pentium processor and Windows 3.1/3.1.1. You will also need a 16-bit sound card with appropriate software drivers. Macintosh versions require a Power PC with MacTCP, a Macintosh Power PC with Open Transport, or a Macintosh 68040 running Apple System 7.5, 7.1, or 7.0. If you have System 7.0, you will need QuickTime Version 1.6.1 or later. Macintosh computers also require 16-bit sound. In addition, some sound and music clips require a 28.8 Kbps or faster connection to the Internet.

RealAudio works by having a series of special servers connected to the Web that each store special RA files. Users locate Web pages using their browsers and click on a RealAudio link. The Web server tells the RealAudio player what audio clip to play and submits a request. The RealAudio server sends the audio stream back to the user and the encoder decompresses the audio. Finally the RealAudio player plays the audio clip. Figure 5-5 shows this procedure.

Let's examine the steps and procedures needed for downloading the RealAudio Player. As you will see, it's quite a simple procedure.

Figure 5-5.
How RealAudio Works

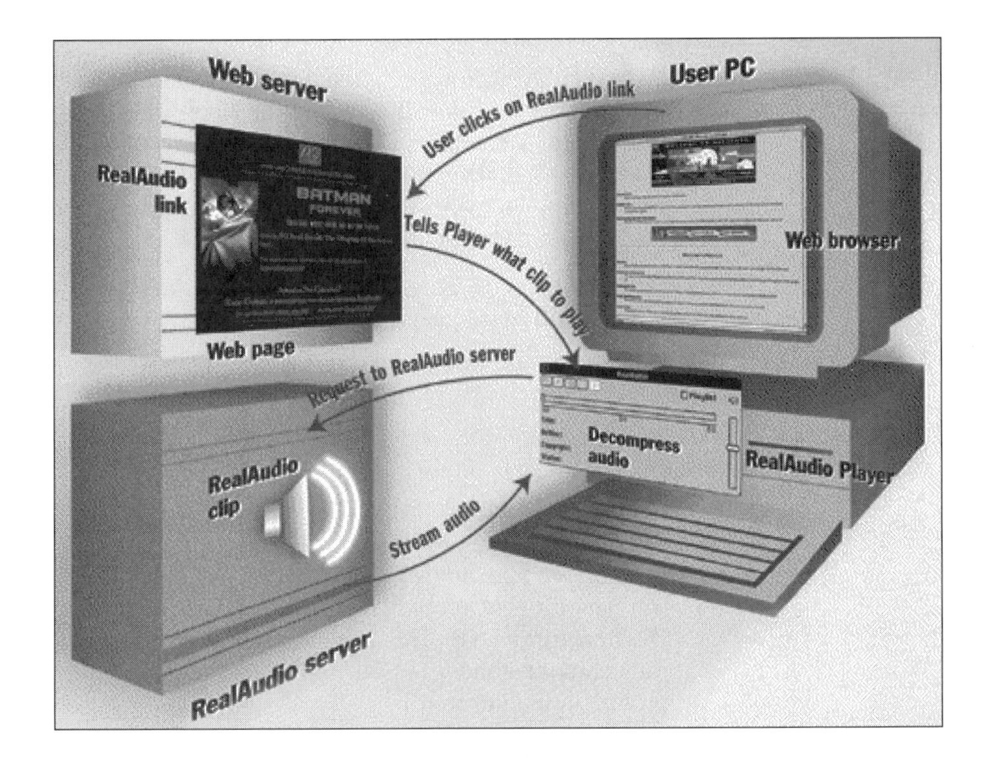

Steps and Procedures for Downloading the RealAudio Player

Follow these steps to download the RealAudio Player for Windows:

1. Access the RealAudio page (http://www.realaudio.com) and click on RealAudio Player and Download Now on the new page. Select RealAudio Player for the product, the correct version of Windows for the platform, your processor type, and your connection speed. Enter your name and e-mail address and click on the Download/Instructions button. On the new page, click to link to one of the sites. Save the file in a temporary sub-directory of your hard drive.

2. Next, you will need to run the downloaded file using one of the following methods: in Program Manager, pull down the File menu, choose Run, and use Browse to locate the downloaded file; or find RA*.EXE using the File Manager and double-click on it to initiate installation. If you have Windows 95 go to the Start menu and choose Run, then use Browse to find the downloaded file.

3. The RealAudio Player Setup window will open with the licensing agreement displayed. You must agree to the license before the Player is installed. Next, you will be asked for your name, company, and connection speed. After this, you will be given the option to select Express or Custom Setup. Select the Express Setup to allow the RealAudio Player to install automatically. If you want to control where and how the player is installed, use Custom Setup.
4. When installation is complete, the RealAudio Player will pop up and announce that the player is installed.

To download the RealAudio Player for the Macintosh computer, follow these steps:

1. Click on RealAudio Player on the Web page (http://www.realaudio.com). Click on RealAudio Player and Download Now on the new page. Select RealAudio Player for the product, Mac OS as the platform, Power PC or 680x0 Mac as the processor type, and your connection speed. Enter your name and e-mail address and click on the Download Instructional button. The file that you will download is a self-extracting StuffIt archive in **BinHex** format. (BinHex is a method of encoding binary files so that they contain standard characters and can be transferred to other computers on the Internet.)
2. If your Web browser is configured for automatic download, a helper application will translate the file and create a RealAudio Player folder containing the Player and a ReadMe file with further instructions. If your browser is not configured for automatic download, click the link, then choose Save To Disk... when prompted by your browser. Use StuffIt Expander to decompress the file. If the download is not automatic and you are not prompted to save the file, you may need to configure your browser for MIME type applications/mac-binhex-40 to Save To Disk or Prompt User. If you would like instructions on downloading the StuffIt Expander via ftp, open the following page using the URL: http://www.realaudio.com/tech_notes/stuffit_help.html. With download accomplished, you're ready to use the RealAudio Player.

Steps and Procedures for Using the RealAudio Player

Click on any RealAudio link on a Web page. The Player will automatically open and audio will play. The Player window will display the author, title, and copyright information about the audio clip. The button on the top lefthand corner of the Player allows you to start and pause audio playback. The Stop button is to the immediate right of the Play button. The view window to the right of these buttons shows the section of the audio clip currently playing. The button to the far right acts like a fast forward and rewind button, allowing you to skip forward or backward in the clip in increments set up in the Seek Time option of the Preferences window. You can also position the slider forward or backward to move around in the clip. Volume is increased by pushing upward on the slider that is located on the right side of the Player. Refer to Figure 5-6.

Now follow these steps to link to, for example, C-SPAN's RealAudio Server:

1. Using Netscape, open the URL http://www.c-span.org/realaudi.html and click on the C-SPAN Weekly Radio Journal graphic at the bottom of the page.
2. Click on Click Here at the left to listen to the full C-SPAN Weekly Radio Journal. If you properly downloaded and configured the RealAudio Player, an RA file should download and the RealAudio Player should appear, as shown in Figure 5-7. You should be able to hear the Weekly Radio Journal broadcast through your computer speakers.

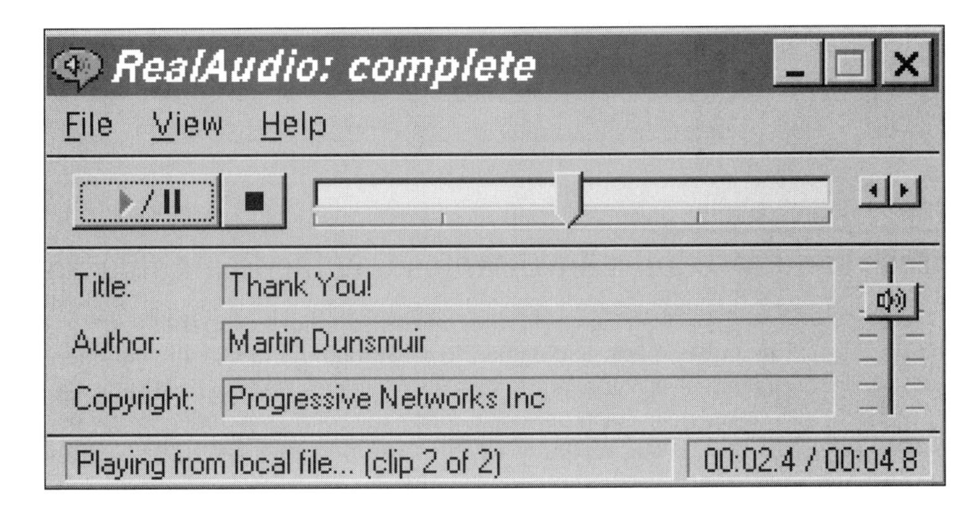

Figure 5-6. The RealAudio Player

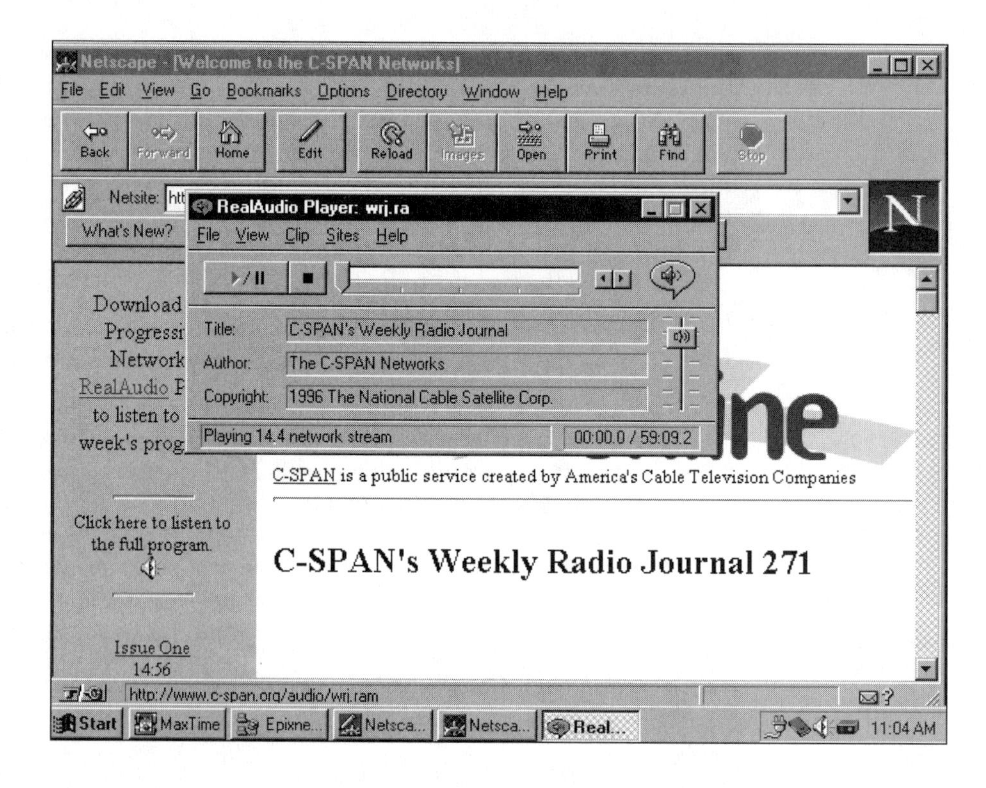

Figure 5-7. RealAudio Player Playing C-SPAN Weekly Radio Journal

Steps and Procedures for Downloading the RealAudio Encoder

To convert an audio file to RealAudio format so that you can include the files on a Web page, you'll need to download the RealAudio encoder. The encoder compresses digital audio files and converts them to RealAudio format. The Windows encoder is capable of reading the following audio file types: WAV, RA, AU, and PCM. The resulting RA files can be played with the RealAudio Player. Follow these instructions for Windows:

1. Click the link on the word Download on the encoder download page (http://www.realaudio.com/products/encoder/index.html). Enter your name and e-mail address, select RealAudio Encoder as the desired product, the correct version of Windows, your processor type, and your connection speed. Click on the Download/Instructions button. Click on Click Here to Download on the new page. Save the file in a temporary subdirectory of your hard disk.

2. The encoder file is a self-extracting archive. Run the downloaded file. Doing this will decompress the Encoder files, including README.WRI, which contains further instructions.

To download and install the RealAudio Encoder for the Macintosh computer, follow these steps:

1. Click on the word Download, enter your name and e-mail address, and select RealAudio Encoder as the desired product, Mac OS as the platform, and Power PC or 680x0 Mac as the processor type from the download page (http://www.realaudio.com/ products/encoder/index.html). The resulting file is a self-extracting StuffIt archive in BinHex format.
2. If your Web browser is configured for automatic download, a helper application will translate the file and create a RealAudio Encoder folder containing the Encoder and a ReadMe file with further instructions. If your browser is not configured for automatic download, click the link, then choose Save To Disk... when prompted by your browser. Use StuffIt Expander to decompress the file. If the download is not automatic and you are not prompted to save the file, you may need to configure your browser for MIME type applications/mac-binhex-40 to Save To Disk or Prompt User. If you would like instructions on downloading the StuffIt Expander via ftp, open the following page using the URL http://www.realaudio.com/tech_notes/stuffit_help.html.

Steps and Procedures for Using the RealAudio Encoder

The following steps will show you how to use the RealAudio Encoder with an IBM or IBM PC–compatible computer. Macintosh users should use a similar procedure with different paths and filenames.

1. Using Windows, load the RealAudio Encoder.
2. Key in the following path under Source on the left side of the encoder window: for Windows 3.1, C:\WINDOWS\CHORD.WAV and for Windows 95, C:\WINDOWS\MEDIA\CHORD.WAV.
3. Place a diskette in the A or B drives of your computer. Replace the file path that appears under Destination with A:\CHORD.RA (or B:\CHORD.RA).
4. Click on Start Encoding in the Encoding menu (or click on the Start Encoding icon on the toolbar). The encoding process should begin and the file should be saved on your designated disk. Compare your screen with Figure 5-8.
5. Load the RealAudio player and play the CHORD.RA file to test it. This file can now be placed on a Web server and played.

Figure 5-8. RealAudio Encoder with an Encoded File

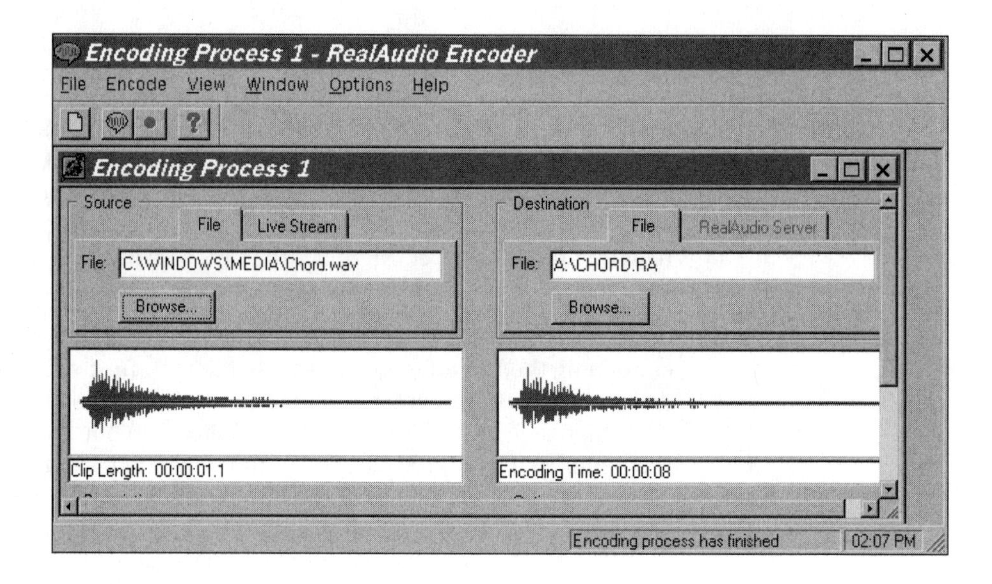

Steps and Procedures for Downloading the TrueSpeech Player

Some Web sites use another real-time audio player, called TrueSpeech. Unlike RealAudio, TrueSpeech (by DSP Group, Inc.) reads TA and WAVE files rather than special RA files created especially for RealAudio. The player system does not require downloading an encoder and therefore the procedures for downloading and configuring it are much simpler than for RealAudio. The player works well and many more sites on the Internet are likely to offer TrueSpeech links in the future. Follow these steps for downloading TrueSpeech for use with Windows 3.1, Windows 95, Windows NT, Macintosh, and the PowerMac.

1. Use Netscape to open the DSP Group page with the following URL: http://www.dspg.com/internet.htm.
2. Click on Download TrueSpeech Player. Click on the desired version of TrueSpeech on the new page.
3. Enter the information requested under the license agreement on the new page. Click on the Download button and choose a temporary subdirectory when prompted.
4. From the File Manager or Explorer, double-click on the file just downloaded. Doing this will self-extract all the necessary files required for installation in step 5.
5. From the File Manager or Explorer, double-click the SETUP.EXE file in the temporary subdirectory. Doing this should initiate the install procedures.

Follow the directions that appear on the screen. The temporary subdirectory can be deleted.

Now you're ready to use the TrueSpeech player.

Steps and Procedures for Using TrueSpeech

To use TrueSpeech, you can either load the player and then connect to a URL by clicking an insertion point in the white text block (see Figure 5-9) or you can access a Web site that contains a file that is TrueSpeech compatible, click on the filename, and spurn the TrueSpeech player.

Follow these steps to use the TrueSpeech Player to play an audio file on a Reflections–Leon Malinofsky Web site.

1. With Netscape, open the following Web page using the URL http://www.crocker.com/~lwm/testdex.html.
2. Click on Voices, which is one of the topics of interest.
3. Click on the TrueSpeech version (marked with a TS) of "The World According to Marcus Aurelius." You may be prompted to configure the browser for TrueSpeech, in which case you must indicate the subdirectory on your hard disk in which the player is located. The player should appear and the selection should begin playing (see Figure 5-10).

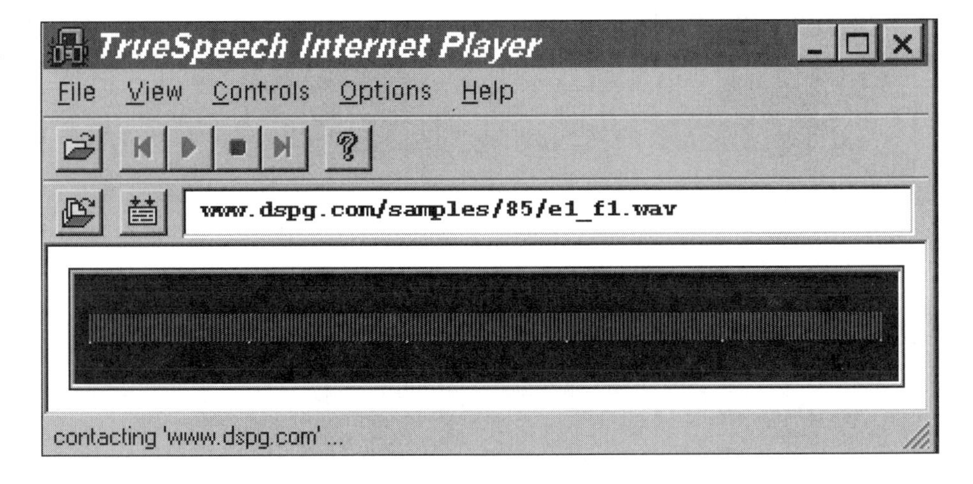

Figure 5-9. TrueSpeech Player with a Keyed File

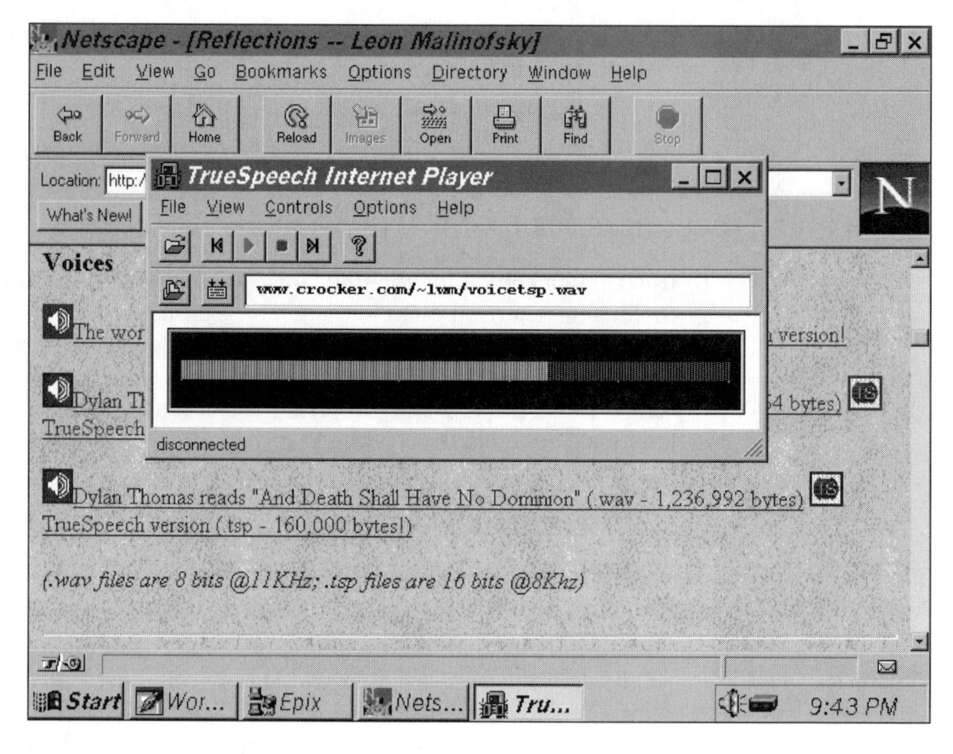

Figure 5-10. TrueSpeech Player Playing a Selection from a Web Page

Questions for Review

1. What are some benefits of using sound?
2. What are some applications for which sound can be used?
3. What are the five main types of audio file formats?
4. How can a PC-compatible computer be configured for sound without using a sound card?
5. Why is an audio helper application an important consideration?
6. What are the names of the popular audio helper applications for the PC and for the Macintosh?
7. What is a real-time audio player? What are the names of two popular real-time audio players?
8. How do you download and install RealAudio and TrueSpeech?
9. How do you use RealAudio and TrueSpeech?
10. What is the purpose of the RealAudio Encoder? How do you use it?

Exercises

1. Download and install one or more of the audio helper applications for the PC-compatible or Macintosh computers. Refer to the text for the necessary URLs. If you are using a PC, you can find information and links to the audio helper applications at http://burgoyne.com/vaudio/net-sound.html.
2. If you are using a PC and it does not have a sound card, download and install the speak.exe file and configure it so that sound files will be played through your PC's internal speaker. Refer to the text for the necessary URLs.
3. Use Netscape to access any of the following Web sites and download and play the audio files you find. You may want to try the sites listed in Table 5-2 in this chapter or one of the search engines to search for additional sites using the key words Sound, Audio, or Music.

   Fun Sounds
   http://www.wb39.com/downsounds.html

   Sounds Directory
   http://www.acm.uiuc.edu/rml/Sounds

   Lemming's Sounds
   http://web.ukonline.co.uk/Members/francesca.obrien/sounds.htm
4. Use Netscape to access the RealAudio homepage (http://www.realaudio.com). Download the current version of the RealAudio player and configure it for use with the browser.
5. After downloading and configuring the RealAudio player, open the RealAudio TimeCast page (http://www.timecast.com). Select a radio or TV program listed under "Live Stations" under the Live Guide listing and listen to the program.
6. After downloading and configuring the RealAudio player, click on one or more of the sites that contain RealAudio files on the RealAudio TimeCast page (http://www.timecast.com) and select one of the sites listed under "Daily Briefing." Select a site that interests you.
7. Download and install the RealAudio Encoder at http://www.realaudio.com/products/encoder/index.html.
8. After downloading and installing the RealAudio Encoder, encode the TADA.WAV file from the WINDOWS subdirectory or another file that you downloaded from an Internet site. Use the RealAudio Player to test the file.

9. Use Netscape to access the TrueSpeech homepage (http://www.dspg.com/internet.htm). Download the current version of the TrueSpeech player and configure it for use with the browser.
10. After downloading and configuring the TrueSpeech player, click on one or more of the sites that contain TrueSpeech files on the TrueSpeech Cool Sites page (http://www.dspg.com/cool.htm) and select one site. Find a TrueSpeech file to play.

CHAPTER

6

Video and Animation on the Internet

This chapter will explain how video and animation are used on the World Wide Web as an additional element of multimedia. It will discuss video/animation file formats, Netscape LiveVideo, video/animation helper application utilities, and the Java programming language along with an example of JavaScript.

What You Will Learn

- How video/animation is used
- Types of video/animation file formats
- How animated GIFs are defined
- How to use Netscape LiveVideo
- Types of video/animation helper applications
- Characteristics of the Java programming language
- How to use Java applets
- How JavaScript is defined

Using Video/Animation

Motion enhances communication. **Video**, which can consist of full-motion video or animation, adds a unique dimension to a Web page and is often the best method for communication. Video is often combined with audio to create a visual effect with a spoken explanation. A **talking head** is a technique in which an onscreen of a person "talks" to the user on a one-to-one basis. Almost everyone in our society is accustomed to this technique from television and video educational and training materials. It makes the presentation much more personal and is a very effective attention getter.

Another video technique incorporates some type of motion. For example, if an auto manufacturer wants to show the public how workers check to see that component parts are installed properly, a still graphic image would not have nearly as much impact as a video of this quality control technique taking place on the production line. Video clips showing leisure time activities and sports such as skiing, basketball, or football are also much more effective than a still photographic image.

Animation, line art that simulates motion, is often used to demonstrate a specific procedure or process, but it does not lend itself to a full-motion video clip. Application examples include a diagram showing a working human heart, the movement of a piece of mail from its origin to destination, and a manufacturing process. Sound can be a part of the animation; on the Web, however, audio is often not included.

Using video and animation on a Web page presents several problems at the present time. One is that the quality is not as good as in video and animation on television or in motion pictures. The video image is limited in terms of resolution and size, and the motion is often somewhat jerky. The problem is that the required bandwidth or speed at which the moving images are transmitted to a user's computer is less than the bandwidth of existing telephone networks.

One alternative is to use an **integrated services digital network** (**ISDN**) connection that can transfer data starting at 56 KB to 128 KB (see Chapter 1). With this type of network phone connection, you can get better but not ideal video quality. ISDN, while an alternative for larger companies, is presently too costly for smaller firms and individuals. After buying a required add-in card for a computer to support ISDN, the user must also consider the fairly expensive ISDN line costs, though the price will probably go down in the future.

A second alternative is the **digital simultaneous voice data** (**DSVD**) technology now being developed. Bill Gates, chairperson and chief executive officer of the Microsoft Corporation, predicts that DSVD technology will be widely adopted over the next three years as an alternative to ISDN. The DSVD technology will greatly improve Web users' ability to download and play full-motion video files.

Television cable companies are eager to enter the Internet access market, and the federal Telecommunications Bill of 1996 opens the door to this possibility. Internet access via television cable will also serve as an alternative to ISDN. The Tele-Communications, Inc., Time Warner, and Viacom cable television companies are investing billions of dollars to upgrade their TV networks to handle the two-way traffic needed to deliver the Internet.

A special television cable modem is required for these computer connections to the Internet. One is the SurfBoard cable modem from General Instruments, and other companies are also developing cable modems. These modems have the potential of delivering more than 10 million bits per second (10 Mbps) to your personal computer. In comparison, a file that takes 10 minutes to download with a 28.8 Kbps modem will arrive in less than 30 seconds with a cable modem. This capability may make it practical to send high-resolution video over the Internet.

Video/Animation File Formats

Four major video/animation file formats are presently used for video/animation on Web pages: Quicktime Cross-Platform (MOV), Audio Video Interleave (AVI), Autodesk Flick (FLI), and the Moving Picture Experts Group Digital-video Standard format (MPG), often called simply MPEG. Each of these platforms has its own system for compressing and storing digital video and each format is *scalable*, meaning that the video will play on any computer even if the original movie was created on a high-end machine at 30 or more frames per second. Table 6-1 provides a listing of the video/animation file formats.

TABLE 6-1 VIDEO/ANIMATION FILE FORMATS

Format	Description
MOV	Quicktime cross-platform
AVI	Audio video interleave
FLI	Autodesk flick format
MPG	Motion Picture Experts Group digital video standard (MPEG)

Quicktime. The Quicktime Cross-Platform video file format (MOV) was originally devised for use with the Macintosh computer long before anything similar was developed for Windows computers. One of its big advantages is that you can now play the files on PC-compatible computers by using Quicktime for Windows.

Quicktime provides an open architecture so that it can play video stored in a variety of compression formats, called **codecs**. It allows users to make technical decisions about frame rates, the amount of compression to be applied, and image quality. The standard interface also lets users control recording and playback.

Audio Video Interleave. The Audio Video Interleave (AVI) format is the standard for Microsoft Video for Windows (VFW). VFW is an add-on for Windows 3.1 but is built in as a part of Windows 95.

Flick. The Autodesk flick format (FLI) are files created with the Autodesk Animator software package. Although FLI files can be used for animation images, they are most often used for cartoon-style animation similar to the type of animation that can also be created with the Java programming language, to be discussed later in this chapter.

MPEG. The **Moving Picture Experts Group Digital-video Standard format** (**MPG**), sometimes known as MPEG, has become the de facto standard for digital video transmitted on the Web. There are two MPG formats—MPEG-1 and MPEG-2. The MPEG-2 format includes extensions to cover a wider range of applications. Its primary application is in transmitting broadcast TV–quality video, but it is also efficient for other applications. Work has begun on new versions of the MPG format.

Many shareware helper application programs will play MPG files, but some cannot encode MPG sound along with the images. Another option for the person who desires to download and play numerous MPG files is to purchase one of the new MPEG boards that can be installed in the PC or Macintosh. The advantages of using a board is that it accelerates the process of encoding the video image and makes sound interleaving possible.

Animated GIF Files

The GIF format's latest version (89a) lets users create simple animation. Browsers like Netscape that support the GIF89a format will show the graphic as a moving image. If the browser doesn't support this format, the image will appear without any animation. Animated GIF files allow multiple images to be compiled within a single GIF file. This "stream" of images can be used like frames in an animation sequence. This means that the single GIF file that is referenced in an HTML file will display multiple images, in sequence, much like a flip-book animation. A good reference for explaining the technical aspects of how animated GIF files work and are created can be found at CyberNet International's "Gif Animation on the WWW" at http://members.aol.com/royalef/gifanim.htm. This site also contains a gallery of sites that contain animated GIF files.

The advantages of using an animated GIF file are:

1. They possess all of the benefits of normal GIF files: transparency, compression, and interlacing for optimum size and compression.
2. They are supported by Netscape with no plug-ins or additional software required.
3. A simple program is all that is required to make animated GIF files.
4. The animation is repeatable and reusable.
5. The animation only loads once, so your modem doesn't need to download constantly.

6. The animations are compact.
7. They work like any other GIF file by including an IMG or FIG tag.

Animated GIF files have the following limitations:

1. They do not work with some browsers.
2. They will only play once or continuously.
3. They will not work as a background GIF and only the first frame will display.
4. The animation may be slowed down or interrupted by other images being downloaded and other playing animations.

The Microsoft GIF Animator program now allows you to animate any GIF file or any image you can copy to the Windows clipboard. It is available for free download by linking to ftp://ftp1.microsoft.com/msdownload/gifanimator.

Many servers on the Internet contain video and animation files. Table 6-2 provides a list of some sample Internet sites for sound files.

TABLE 6-2 SAMPLE INTERNET SITES FOR VIDEO/ANIMATION FILES

Site Name	URL
CNN Video Vault	http://www.cnn.com/video_vault/index.html
Dream Video Archive	http://www.crl.com/~dreamnet/avclips.html
Thant's Animation Index	http://mambo.ucsc.edu/psl/thant/thant.html
MPEG Archive	http://www.powerweb.de/mpeg/
Destiny's Animated GIF Collection	http://www.geocities.com/siliconvalley/Park/2100/

Video and Netscape

Netscape 3.0 contains a feature that allows users to instantly view AVI movies either embedded in or linked to Web pages. With this feature you are able to click on a movie image to play it and click again to stop. A right mouse click on an image pops up a complete menu of controls including Play, Pause, Rewind, Fast Forward, Frame Back, and Frame Forward. This feature makes it unnecessary to download a separate sound helper application to play AVI sound files.

You may be using an earlier version of Netscape or some other browser that needs an AVI video file helper application. Next, we'll examine some of the popular video helper applications for both the IBM or IBM PC–compatible and Macintosh computers.

Video/Animation Helper Applications

To be able to play video/animation on the Web using both IBM or IBM PC–compatible or Macintosh computers and a browser like Netscape, you need a video/animation helper application program, which is usually not built in. Fortunately, various helper applications are available on the Internet FTP servers. These video/animation application programs execute when a supported video/animation file is downloaded from the Web. You may need to click on the Play button (usually a right arrow) on most helper applications when they appear. Unlike the helper applications for sound, video/animation helper application programs can usually play only one type of video/animation file. The following are the most frequently used video/animation helper application programs for both the IBM or IBM PC–compatible and Macintosh computers.

IBM/IBM PC-COMPATIBLE COMPUTERS

A helper application for Quicktime (MOV) files is the Windows Quicktime Player, which was created at the Chinese University in Hong Kong. Windows 95 should already contain the Quicktime Player. (If you're using Windows 3.1, you'll need to download it.) The player works in conjunction with the Media Player and lets you play Quicktime movies. You can download it at ftp://ftp.cuhk.hk/pub/mov/qtw111.exe.

After you download the file, follow these steps to install the player: (1) Execute the qtw111.exe file at the DOS prompt. The file will expand into three files: DISK1.ZIP, DISK2.ZIP, and README.TXT. (2) Unzip the two ZIP files onto two separate floppy disks. (3) In Windows, run SETUP.EXE from the floppy DISK1 by choosing Run from the File menu. You must run SETUP.EXE from floppy disk, or the following error will occur: "The destination file xxxx has different number of bytes as the source file xxxx. Please ...." (4) Follow the onscreen instructions for installing the software. The file SAMPLE.MOV is included in the installation and you can use it to test the player. For both Windows 3.1 and Windows 95, you will need to configure the Quicktime player when you encounter a MOV file in Netscape. Indicate the path for the Quicktime Player as C:/QTW/BIN/PLAYER.EXE. Substitute another drive letter if your hard disk is not C. Figure 6-1 shows the Quicktime Player.

If you're using Windows 95, Audio Video Interleave (AVI) files will play through the Media Player (MPLAYER.EXE) which comes packaged with this version of Windows. If you're using Windows 3.1, you'll need to download the Microsoft Video for Windows Viewer helper application. The URL for this download is http://www.microsoft.com/kb/softlib/mslfiles/wv1160.exe.

The Autodesk flick format (FLI) animation files require a downloaded player for Windows 3.1, but the files will also play through the Media Player with Windows 95. The Autodesk Animation Player may be downloaded at ftp://ssp.hk.super.net/.3/cica/win3/desktop/waaplay.zip. Figure 6-2 shows the Autodesk Animation Player with a loaded FLI file.

Figure 6-1. The Quicktime Player

Figure 6-2. The Autodesk Animation Player

The Moving Picture Experts Group Digital-video Standard format (MPG) files may require helper applications for both Windows 3.1 and Windows 95.

MACINTOSH

Two Macintosh video helper applications are available. One is Sparkle, which plays and converts both MPEG and Quicktime files. It is available for downloading at ftp://www-dsed.llnl.gov/files/programs/MAC/. Another application, SimplePlayer, plays Quicktime files and is available at http://quicktime.apple.com. A helper application called Fli-Viewer will play Autodesk flick animation files and is available at ftp://crusty.er.usgs.gov/pub/animation/fli/. The MacAmin helper application will also play flick files. It is also available for download at ftp://crusty.er.usgs.gov/pub/animation/fli/.

The Java Programming Language

Have you heard about Java? Sun Microsystems, Inc., which designed this famous programming language, describes **Java** as "a simple, object-oriented, distributed, interpreted, robust, secure, architecture-neutral, portable, high-performance, multithreaded, and dynamic language." This language provides another method to incorporate animation and sound into Web pages by using either Netscape 2.0 and above or the HotJava browser. Other Web browsers are now Java compliant. The following are Java's eleven main features:

1. **Simple.** Java can be easily programmed without much training, using the C++ language as its basis. However, the developers omitted those C++ features that might be difficult for nonexperts to understand. Java also requires a small amount of programming language and can run standalone on small machines. The size of the basic interpreter and class support is about 40 KB.

2. **Object-oriented.** This technique focuses design on the data and interfaces to it. Object-oriented programming is independent enough to stand on its own and can be copied into other programs. Java's object-oriented programming is facilitated by using C++.

3. **Distributed.** This feature means that Java routines for coping with various TCP/IP protocols are available for downloading from Sun Microsystems and from other places on the Internet. These routines are provided free from the developers of Java and other programmers working with the language.

4. **Interpreted.** The Java interpreter can execute Java codes directly on any machine to which the interpreter has been ported. Because of this feature, the development process becomes much more rapid and exploratory.

5. **Robust.** Java allows programmers the opportunity to make choices explicitly as the code is being written. If something is done incorrectly, Java will inform you about it when it is compiled.
6. **Secure.** Java applications are guaranteed to be virus free and tamper resistant because they cannot access system memory. Java uses public-key encryption to effectively prevent hackers from examining protected pieces of information such as account names and passwords.
7. **Architecture-neutral.** This means that Java-compiled code is executable on many different network processors connected to the Internet with the Java runtime system. Java will run on IBM or IBM PC–compatible and Macintosh computers. This is an important feature for application writers and for individuals who desire to use Java when writing HTML documents.
8. **Portable.** Java libraries define portable interfaces so that Java can be easily moved from one system to another.
9. **High-performance.** There are situations in which higher performance is required and Java is written to meet these requirements.
10. **Multithreaded.** A multithreaded program can run two or more independent portions of the program (threads) at the same time. Multithreading allows Java users to experience better interactive responsiveness and real-time behavior. However, when Java runs on top of other systems like UNIX, Windows, and the Macintosh, real-time responsiveness is limited.
11. **Dynamic.** Java was designed to adapt to a evolving environment.

COMPLAINTS ABOUT JAVA

All its virtues notwithstanding, there have been criticisms of Java among some Web page developers. Let's look at what the critics have to say. Developers' complaints about Java seem to focus on three areas. (1) There is generally a lack of good documentation. Recently, however, the documentation available online seems to have greatly improved. (2) Java is not available for every user. In the past, anyone wanting to work with Java needed a high-end machine such as SparcStations and Pentium PCs. (3) Although Java's architecture is platform independent, its runtime system is somewhat slow. This complaint should evaporate when Sun Microsystem's new inline native compiler is installed.

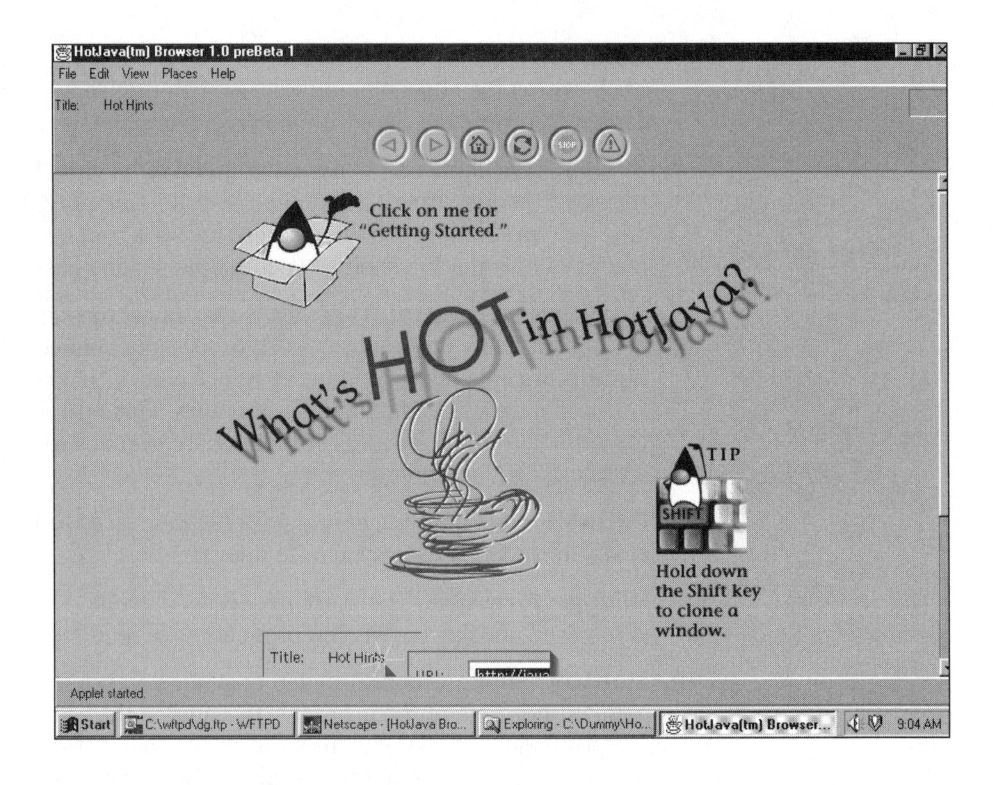

Figure 6-3. The HotJava Browser

HotJava

HotJava, the browser specifically designed to be used with Java, can be substituted for another browser. Developers claim that some types of Java "behavior" can only be viewed with HotJava. At the time this textbook was written, HotJava was only available for Windows 95 and Windows NT. HotJava can be downloaded at http://java.sun.com/HotJava/index.html. Figure 6-3 shows the HotJava browser.

Using Java Applets

The two ways to use Java are by writing a standalone application or by using a Java **applet**, a program written in Java programming language that can be included in an HTML page, much like including a graphic image. When you use Netscape to view a page that contains a Java applet, the applet's code is transferred to your computer and is executed by the browser. At the time that this textbook was written, sample Java applets were available but would only run on PC-compatible computers using Windows 95 and Windows NT. Sample applets are available for viewing and downloading from http://java.

sun.com/applets/applets.html and http://www.gamelan.com. Figure 6-4 shows the animator applet and the character "Duke." Sound also accompanies this applet.

The HTML codes that support the applet shown in Figure 6-4 are as follows:

```
<title>The Animator Applet</title> <hr> <applet code=Animator class width=200
height=200> <param name=imagesource value="images/Duke"> <param
name=endimage value=10> <param name=soundsource value="audio"> <param
name=sound track value=spacemusic.au> <param
name=soundsvalue="1.au|2.au|3.au|4.au|5.au|6.au|7.au|8.au|9.au|0.au">
<param name=pause value=200>
</applet>
<hr>
<hr>
<a href="Animator.java">The source</a>
```

There are two ways to embed Java applets into HTML documents. The first way is to use a local applet, a type of applet that is retrieved from your local system. The second way is to use a remote applet, which is loaded from a remote server on the Internet. To use a remote applet, you need to know the URL for the applet. This location on the remote server contains the code needed to have the applet function properly for you. You also need to know the set of optional and required attributes specific to the applet. These attributes should be published on the server by the applet's author(s).

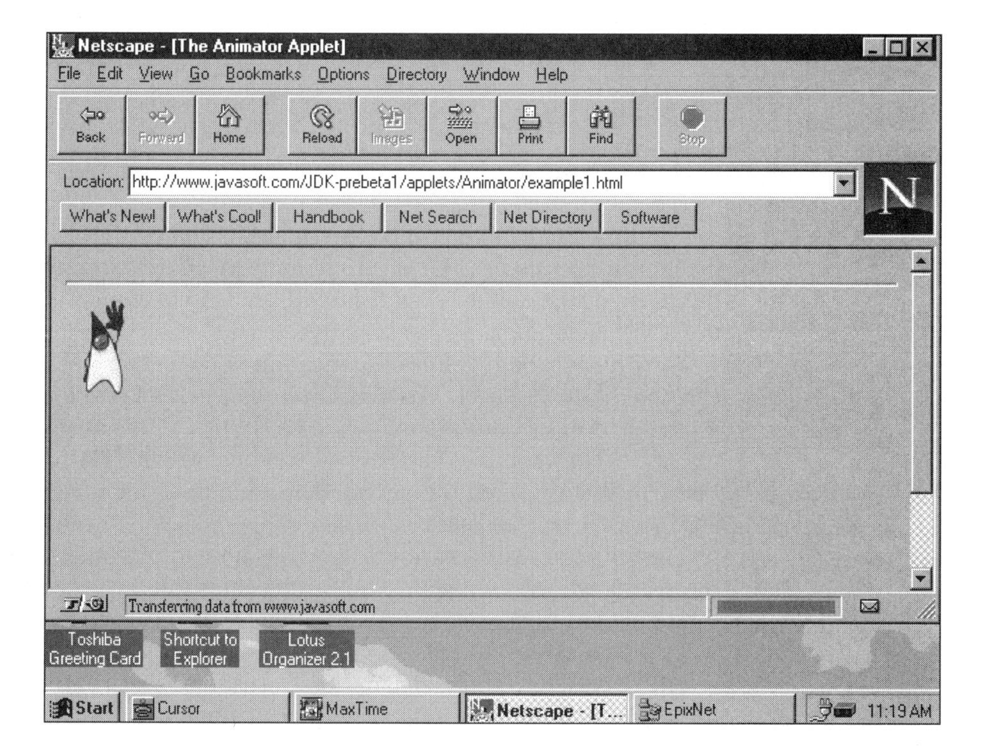

Figure 6-4. "Duke" and the Animator Java Applet

ADDING A JAVA APPLET

To add a Java applet, the APP HTML element is used. The APP element follows this syntax for several required and optional attributes. Most HTML editors don't yet support JAVA, so the attributes will need to be keyed in manually.

```
<APP
   CLASS="class name"
   SRC="URL"
   ALIGN="alignment"
   HEIGHT="height in pixels"
   WIDTH="width in pixels"
   APPLET_SPECIFIC_ATTRIBUTES="values"
...>
```

Here is a specific example of an HTML document with an actual workable embedded Java applet for use with the HotJava browser. If the filename for this HTML document is appex1.htm and it is saved in the C:\SAMPLES subdirectory, you would key in file:///c:/samples/appex1.htm to open the document.

```
<HTML>
<HEAD>
<TITLE>An Applet Example</TITLE></HEAD>
<BODY>
<APP CLASS="ImageLoopItem"
   SRC="doc:/demo/"
   IMG="doc:/demo/images/duke"
   WIDTH=55
   HEIGHT=68>
</BODY>
</HTML>
```

The CLASS attribute, which is required, defines the name of the applet to invoke the location in the HTML document. In the example, ImageLoopItem is the name of the applet. Exact spelling and capitalization is necessary because the UNIX file system is case sensitive.

The SRC attribute defines the location in the server's system where HotJava can find the code object. In the example, doc:/demo/ is the location of this applet. This is an optional but recommended attribute. If this attribute is not specified, HotJava will search the directory where the current HTML document is located. You can specify a remote server that contains applets you want to use (i.e., http://java.sun.com/applets).

The optional ALIGN attribute specifies how to orient an applet in relation to the text flow that follows it. The three values that are used for ALIGN are bottom, top, and middle. If the ALIGN attribute is not specified, the bottom value will be used as the default.

WIDTH and HEIGHT are also optional attributes that control the dimensions of the applet. The values for each of these attributes is designated in pixels. In the example, 55 pixels is specified as the width and 68 pixels as the height. The APPLET_SPECIFIC_ATTRIBUTES can consist of any number of optional additional attributes.

We will now examine an example of a HTML document that uses JavaScript.

JavaScript

JavaScript is the Java scripting language that uses Java to accomplish a specific automated task in a Web document. Using JavaScript, HTML pages can perform the following functions: total orders, react to user input, change the page's appearance, generate documents, and perform other tasks. The following HTML file uses JavaScript tags to change the background color of the Web page and was created by Denise K. Wynn of Loginet, Inc. The document can be accessed with a Netscape or other Java-compliant browser at http://www.loginet.com/users/d/denise/bgcolor/bgcolor.html.

```
!doctype html public "-//IETF//DTD HTML//EN">
<HTML>
<HEAD>
<META NAME="AUTHOR" CONTENT="Denise K Wynn">
<TITLE>JavaScripting Test Page</TITLE>
<script language="Javascript">
<!—
 function backcolor(form){
    temp = ""
    for (var i = 0; i < 16; i++) {
               temp = form.color[i].value
               if (form.color[i].checked){ document.bgColor = temp }
               }
    }
function randombackground(){
    document.bgColor = getColor()
    }
function getColor(){
    currentdate = new Date()
    backgroundcolor = currentdate.getSeconds()
    if (backgroundcolor > 44)
               backgroundcolor = backgroundcolor - 45
    else if (backgroundcolor > 29)
               backgroundcolor = backgroundcolor - 30
    else if (backgroundcolor > 15)
               backgroundcolor = backgroundcolor - 16
```

Continued on next page

Continued from previous page

```
    if (backgroundcolor == 0 )
            return "olive";
    else if (backgroundcolor == 1 )
            return "teal";
    else if (backgroundcolor == 2 )
            return "red";
    else if (backgroundcolor == 3 )
            return "blue";
    else if (backgroundcolor == 4 )
            return "maroon";
    else if (backgroundcolor == 5 )
            return "navy";
    else if (backgroundcolor == 6 )
            return "lime";
    else if (backgroundcolor == 7 )
            return "fuschia";
    else if (backgroundcolor == 8 )
            return "green";
    else if (backgroundcolor == 9 )
            return "purple";
    else if (backgroundcolor == 10 )
            return "gray";
    else if (backgroundcolor == 11 )
            return "yellow";
    else if (backgroundcolor == 12 )
            return "aqua";
    else if (backgroundcolor == 13 )
            return "black";
    else if (backgroundcolor == 14 )
            return "white";
    else if (backgroundcolor == 15 )
            return "silver";
    }
  // —>
  </script>
  </HEAD>
  <body onLoad="document.bgColor='lime'; return true;">
  <h3> An obnoxious JavaScript page to show you what you can do with the bgColor
  property.</h3>
  <p>
  <A HREF="" onMouseOver="randombackground()">Choose a random background
  color using onMouseOver</A><p>
  <hr>
  <form>
```

```
Click on the button to choose a random background color<p>
<input type="button" value="Random" onClick="randombackground()">
</form>
<hr>
<form name="backcolorform">
Select a background color.<p>
<table>
<tr>
<td><input type="radio" name="color" value="green"
onClick="backcolor(this.form)">Green</td>
<td><input type="radio" name="color" value="aqua"
onClick="backcolor(this.form)">Aqua</td>
<td><input type="radio" name="color" value="red"
onClick="backcolor(this.form)">Red</td>
<td><input type="radio" name="color" value="olive"
onClick="backcolor(this.form)">Olive</td>
</tr>
<tr>
<td><input type="radio" name="color" value="teal"
onClick="backcolor(this.form)">Teal</td>
<td><input type="radio" name="color" value="blue"
onClick="backcolor(this.form)">Blue</td>
<td><input type="radio" name="color" value="maroon"
onClick="backcolor(this.form)">Maroon</td>
<td><input type="radio" name="color" value="navy"
onClick="backcolor(this.form)">Navy</td>
</tr>
<tr>
<td><input type="radio" name="color" value="gray"
onClick="backcolor(this.form)">Gray</td>
<td><input type="radio" name="color" value="lime"
onClick="backcolor(this.form)">Lime</td>
<td><input type="radio" name="color" value="fuschia"
onClick="backcolor(this.form)">Fuschia</td>
<td><input type="radio" name="color" value="white"
onClick="backcolor(this.form)">White</td>
</tr>
<tr>
<td><input type="radio" name="color" value="purple"
onClick="backcolor(this.form)">Purple</td>
<td><input type="radio" name="color" value="silver"
onClick="backcolor(this.form)">Silver</td>
<td><input type="radio" name="color" value="yellow"
onClick="backcolor(this.form)">Yellow</td>
```

Continued on next page

Continued from previous page

```
<td><input type="radio" name="color" value="black"
onClick="backcolor(this.form)">Black</td>
</tr>
</table>
</form>
<hr>
<script language="Javascript">
<!—
   var theDate = ""
   theDate = document.lastModified
   document.write("Last Modified: ");
   document.write(theDate);
   document.write();
//—>
</script>
<br>
<p>
Created by denise<p>
For questions or information on this page please e-mail<ADDRESS><A
HREF="mailto:denise@loginet.com">denise@loginet.com</a></ADDRESS><p>
<a href="http://www.loginet.com/users/d/denise/javascript.html">[Denise's
JavaScript Pages]</a>
</body>
</html>
```

Questions for Review

1. What are the benefits of using video on a Web page? Provide specific examples.
2. What are the benefits of using animation on a Web page? Provide specific examples.
3. What are the current problems with using video on the World Wide Web? How will these problems probably be resolved in the future?
4. What are the four most frequently used video and animation file formats? What are the characteristics of each format? Which format has become the de facto file format standard for digital video on the Web?
5. What is the difference between MPEG-1 and MPEG-2?
6. What is an animated GIF file? What are the advantages and limitations of using them?
7. What are the most commonly used video and animation helper applications for the PC-compatible computer and for the Macintosh computer?

8. What is the Java programming language? What are its eleven main features?
9. What are some of the complaints about the Java programming language?
10. What are the two ways that Java can be used?
11. What is a Java applet?
12. What are the two ways to embed a Java applet into a HTML document?
13. What are the characteristics of the HotJava browser? Why would it be used instead of a browser like Netscape?
14. What are the purposes and uses for each the following types of attributes in the APP HTML element: CLASS, SRC, ALIGN, HEIGHT, WIDTH, and APPLET_SPECIFIC_ATTRIBUTES?
15. What is JavaScript? Provide an example of how JavaScript is used.

Exercises

1. Download and install one or more of the video and animation helper applications for the PC-compatible or Macintosh computers. Refer to the text for the necessary URLs.
2. Using Netscape to access any of the following Web sites, download and play the video/animation files that you find. You may want to try the sites listed in Table 6-2 in this chapter or one of the search engines to search for additional sites using the key words Video, Animation, or Movies. (Note: You will need to have the appropriate helper application downloaded or installed to play these files.)
   http://www.mcs.net/~comeback/fli/fli.html
   http://www.acm.uiuc.edu:80/rml/Mpeg/
   http://www.disney.com/DisneyPictures/
3. Use Netscape to access "GIF Animation on the WWW" at http://www.reiworld.com/royalef/gifabout.htm. Read the information about how animated GIF files are created. Link to the gallery of sites that contain animated GIF files.
4. Use Netscape to download the current version of the HotJava browser at http://java.sun.com/HotJava/index.html, if one exists that is compatible with your computer and operating system. Install and configure the browser. Select the Java documentation item on the Help menu. Read and study the documentation. Select Cool HotJava Demos on the Welcome page or HotJava Demos on the Help menu. Try one or more of these Java sample pages.

5. Use Netscape to link to Yahoo (http://www.yahoo.com) and select Computers and Internet/Graphics/Computer Animation/Animated GIFs/Collections. Link to several sites that display animated GIF files. Save several GIF files that you like to a floppy diskette.

6. Use the Netscape, HotJava, or other Java-compliant browsers to open one or more of the following Web sites that contain Java applets. Select and play one or more applets that are compatible for Netscape or HotJava.
   http://java.sun.com/applets
   http://www.gamelan.com
   http://www.dimensionx.com
   http://www.servonet.com/javaStuff/Welcome.html

7. Use a HTML editor to key the following HTML page containing a Java applet. Save the file using exer7.htm (exer7 if you are using a Macintosh) as the filename. Try this document with HotJava if you have downloaded and installed it. (Note: This is a sound applet and may only work if your computer has a sound card installed.)

```
<HTML>
<HEAD>
<TITLE>Another Applet Example</TITLE></HEAD>
<BODY>
<APP CLASS="AudioItem"
   SRC="doc:/demo/"
   SND="HELLO.AU">
</BODY>
</HTML>
```

8. Access the JavaScript Resource Center at http://jrc.livesoftware.com for additional examples of JavaScript.

9. Use Netscape to access Denise K. Wynn's JavaScript for changing the background color of a Web page at http://www.loginet.com/users/d/denise/bgcolor/bgcolor.html.

CHAPTER

7

Computer Conferencing on the Internet

Computer conferencing is another exciting multimedia application on the Internet. A thorough explanation of the types of available free or inexpensive computer conferencing applications will be provided. The steps and procedures for using Netscape Chat, WebChat, Internet Phone, and CoolTalk will be given.

What You Will Learn

- How computer conferencing is defined
- Types of computer conferencing application programs
- Steps and procedures for using Netscape Chat
- Steps and procedures for using WebChat
- How telephony computer conferencing is defined
- Steps and procedures for using Internet Phone
- Steps and procedures for using CoolTalk

Computer Conferencing Defined

Computer conferencing is an electronic means of sending, viewing, and sharing real-time communications in areas of common interest via one or more of the following media formats: text, graphics, audio, video, and/or animation. Computer conferencing is one type of teleconferencing, which also includes audio conferencing, videoconferencing, and business television. (Chapter 8 will examine videoconferencing on the Internet.) Computer conferencing allows you to communicate or **chat** via your computer with other people who share the same network connection in real time. That is, you are able to carry on "live" conversations with these individuals on any topic of your choosing. As you key your message, the other users connected see your message on their computer screens. Some computer conferencing systems allow you to include links to graphics and audio files.

In business firms, computer conferencing augments or replaces face-to-face conferences. It can be used, for example, to combine the back-and-forth discussion of a meeting with the capability to retrieve research reports and other documents stored on the network. The online conference gives all employees an opportunity to add suggestions and comments to the discussion. The employee responsible for managing the conference is called a **moderator**. This person establishes the conference, specifies rules of membership, conducts the conference, and terminates it. Computer conferences can be private and public. Private conferences are restricted to a preselected group of people; public conferences are open to anyone.

Here are some of the benefits of computer conferencing over face-to-face conferencing.

- Companies and organizations save the money required to move people to a common site, as well as the time required for travel.
- Participants can easily review the comments made by other participants in the conference.
- Participants can join the conference at any time.
- Participants with a limited amount of time to participate can do so at their convenience.
- A written record of the conference proceedings may be produced.

Computer conferencing has some problems, too.

- The desirable socialization that occurs during a face-to-face conference is lost.
- Equipment required for computer conferencing, such as computers, modems, network connections, and software, may not be available.
- Communication may be slowed because of some individual's poor keying ability.

Types of Computer Conferencing Application Programs

The two main types of computer conference application programs are UNIX-based programs and Web-based programs. The UNIX-based programs are text based and operate with a UNIX server. The Web-based programs are multimedia programs that can use text, graphics, and audio and operate with a Web browser such as Netscape.

UNIX-BASED PROGRAMS

Several text-based computer conferencing programs have been available for use on the Internet for some time. These cannot be considered multimedia applications because they are UNIX-based and don't require using a Web browser such as Netscape. Examining how these programs work, however, will help you to better understand the multimedia-based programs. Ask your network administrator which of the following text-based programs are available for your use. You may have already used these programs, as most local networks have one or both of them available.

The simplest of the text-based programs is called Talk. Talk must be installed on your host computer and both parties must be logged on for a Talk session to occur. One person initiates the session, and the other person accepts or rejects it. To initiate a session, the initiator would key, for example, talk jsmith @grantu.edu at the UNIX prompt. If you are logged on, your terminal will beep and the following message will appear on the screen, asking you if you want to talk: talk: connection requested by ssmith@pluto.henry.edu. If you do, key talk@pluto.henry.edu and the screen will split in two, with the top of the screen displaying what the initiator keys and the bottom half what you key.

Internet Relay Chat. Internet Relay Chat (**IRC**) is the most used text-based computer conferencing program. IRC is either installed on your host computer or you can telnet to another site that has made IRC available. When IRC has been installed on your local network, keying irc (lower case) at the UNIX prompt usually initiates the program. All IRC participants assign themselves nicknames or **handles**, and conversations are conducted on **channels** where people discuss various topics. (A similar system used on multimedia computer conferencing systems will be discussed later.)

After assigning yourself a nickname and logging on to initiate an IRC session, you key /list to see the names of the active channels, numbers of current participants, and the channel topics. After finding a channel that you'd like to join, you key /join #channelname where channelname is the actual name of the channel. You can key your message, which should appear on the screen along with the messages from others who have joined the channel. Some channels require you to have an invitation to join, in which case you key /who #channelname to see a list of the participants. You then pick a name of a person with @ next to their name and send them a message (i.e., /msg name I'd like to join #channelname).

Some IRC channels maintain Web pages. The IRC Channels on WWW page provides an index to the Web pages of over 200 IRC channels, including such channels as #comics, #romance, and #windows95. The typical page has an FAQ for users of the channel, a directory of frequent channel users (often with photos), and a collection of links to related sites.

WEB-BASED PROGRAMS

Several multimedia computer conferencing programs are Web-based for use with the Internet. These programs open up communication possibilities that are not possible or ineffective with the text-based programs. They can create a community atmosphere at your site, add value, and easily promote your site to the public. For example, individuals can use the multimedia computer conferencing programs to share favorite Web sites, provide a tour of the Web to friends or associates, or share graphics and audio files. The following are a few of the many ways businesses can use multimedia computer conferencing programs.

- Live customer product support
- Training sessions
- Communication with resellers and partners
- Customized content programs (e.g., live weekly interviews, presentations, and tutorials)
- Sales presentations and demonstrations
- Live merchandising of products and services
- Internet press conference

The Netscape Chat Program

Netscape Chat is a program available for purchase as a part of Netscape's Power Pack. It is also available as freeware for both IBM and IBM PC–compatible and Macintosh computers. The 16-bit program for Windows 3.1 can be downloaded at ftp://ftp.netscape.com/pub/chat/windows/, and the 32-bit program for Windows 95 can be downloaded at ftp://ftp.netscape.com/pub/chat/windows/. The Macintosh version can also be downloaded at ftp://ftp.netscape.com/pub/chat /mac/.

Netscape Chat is an add-on program that works with the Netscape browser to send, view, and share URLs with other chat users. In addition, URLs can be automatically sent to other chat users, who can immediately view them. Netscape Chat requires a minimum of a 386sx processor, 1 MB of hard disk space, and 8 MB of memory (16 MB recommended). It currently works with Windows 3.1, Windows for Workgroups 3.11, Windows 95, and Windows NT.

Netscape Chat uses multiple chat communication modes: personal conversations (one to one), group conferences (many to many), and moderated auditoriums (one to many). Multiple chat rooms allow users to simultaneously participate in several chat rooms at once. Netscape Chat supports standards-based IRC chat by connecting to any standard IRC chat server, Netscape community system chat servers, and IRC slash commands (explained in the previous section).

Steps and Procedures for Using Netscape Chat

Follow these steps for installing and using Netscape Chat. Make sure that the Netscape browser is in the C:\NETSCAPE directory of your hard disk before you begin.

1. Run the downloaded `nc---.exe` file. (The filename should begin with `nc`.) (The program files will extract.)
2. Install the program by running the `exe` file.
3. Connect to the Internet and double-click on the Netscape Chat icon to load the program. (The Netscape Chat program and the Netscape browser will load and the Server Connection dialog box will appear.)
4. Accept the default server. Enter the following new user information: real name, user name, nickname, password, and electronic mail (e-mail) address. Click on OK. (The connection should be made. If the connection fails, click on the Address button and try one of the other servers listed. See Figure 7-1.)
5. Click on the List button on the Quick Join dialog box, which should appear. (The Conversation Channels dialog box will appear with a list of available channels listed by name of channel, count, and topic. See Figure 7-2.)
6. Select one of the channels by clicking on it and click on the Join button (for example, `#newbies`). (The conferencing dialog box will appear.)
7. Key a message in the white typing box and press the Enter key. (The message should appear in the shaded area preceded by your nickname.)
8. If you know a URL for your school or department's homepage or for your personal homepage, key it in the URL typing box. Click on the Send button. (The URL should appear in the shaded area. Notice that you can also view the URL in Netscape by clicking on the View button. You can also add the URL to a URL list by clicking on the Add button. See Figure 7-3.)

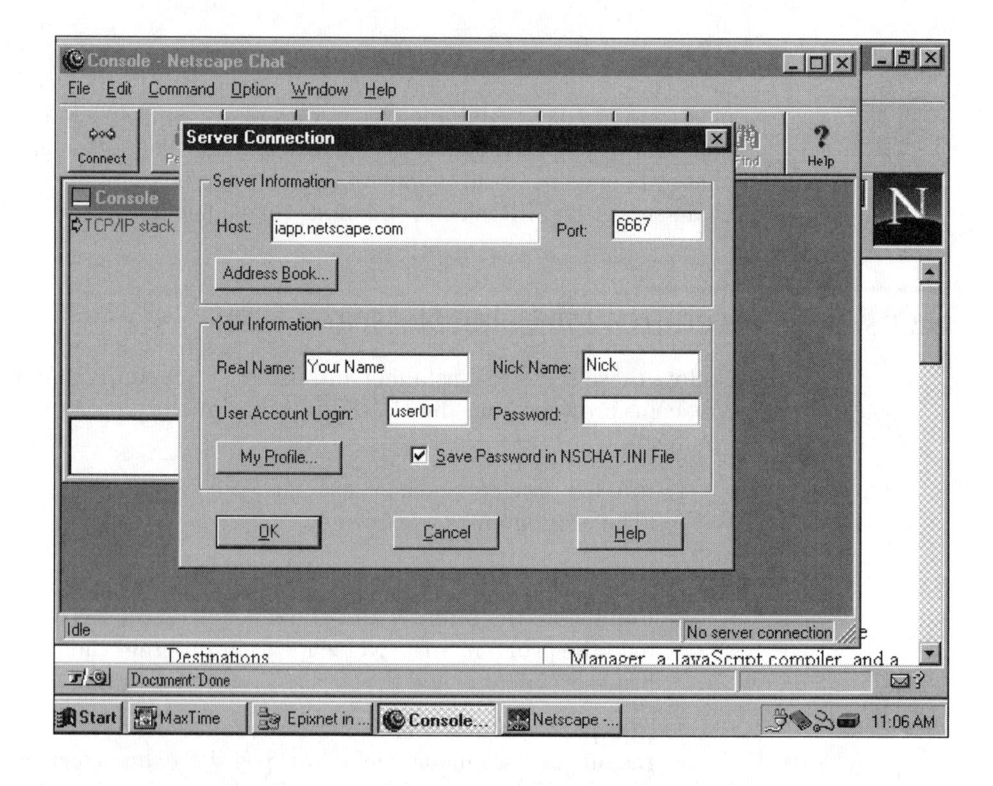

Figure 7-1.
The Netscape Chat Server Connection Dialog Box

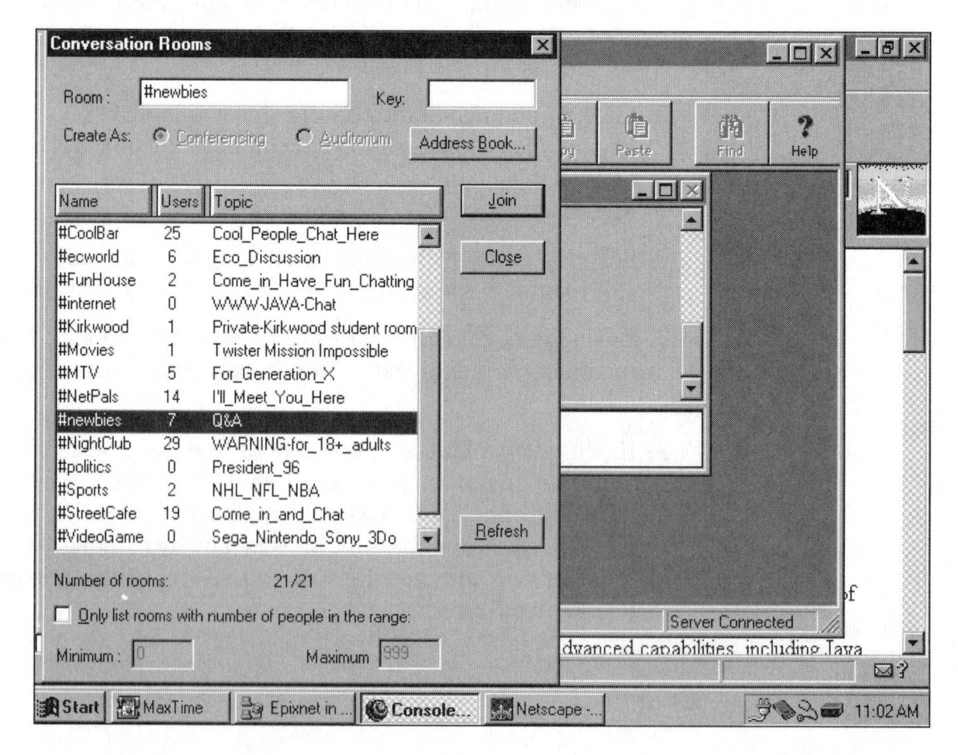

Figure 7-2.
The Netscape Chat Conversation Channels Dialog Box

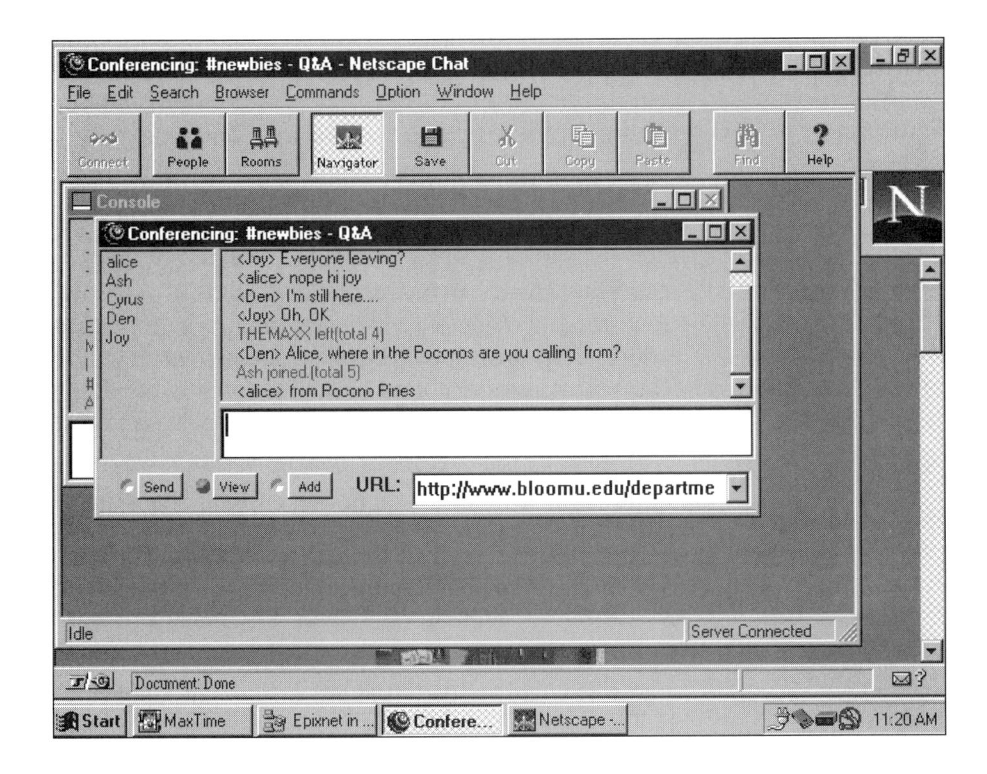

Figure 7-3. The Netscape Chat Conferencing Dialog Box

9. After you are finished sending messages, close the Conferencing dialog box by double-clicking on the control box (-) in the upper left corner if you're using Windows 3.1. If you're using Windows 95, click on the control box (X) in the upper right corner.
10. Click on Disconnect on the File menu.
11. Click on Exit on the File menu to exit the program.
12. Click on Exit on the File menu of Netscape to exit it.

The WebChat Program

WebChat is another very popular, real-time fully multimedia computer conferencing application for the Web. WebChat allows visitors at Web sites to engage in live conversation similar to Netscape Chat. The difference between WebChat and Netscape Chat is that users can quickly incorporate images, video and audio clips, and hotlinks into their chat. WebChat server software is available that runs as an add-on to most Web servers. However, you can use WebChat with any standard browser (including Netscape, Microsoft Explorer, AOL, Prodigy, and Compuserve), and no special software needs to be downloaded to immediately begin chatting. The URL to access WebChat is http://www.irsociety.com/wbs.html.

Steps and Procedures for Using WebChat

Follow these steps to access and use WebChat.

1. Open WebChat with Netscape or another browser, using the preceding URL. The WebChat Broadcasting System Web page should load. (See Figure 7-4.)
2. Click on New Users on either the graphic or hypertexted link. Click on Register and complete the registration form, including your e-mail address. Click on Join WBS! WebChat will send you an e-mail message. Read the message, noting the validation number. Reply to the message to validate your WBS account. Click on OK! I've replied to my validation e-mail.
3. Browse the available Chat hubs, scrolling down the page to see the hub categories.
4. Link to a hub by clicking on a button (i.e., Travel). An introduction to the hub page will appear.
5. Click on a chat room doorway and the Go button. Enter your handle, password, and validation number. Click the Chat Now! button. The wel-

Figure 7-4.
The WebChat Main Page

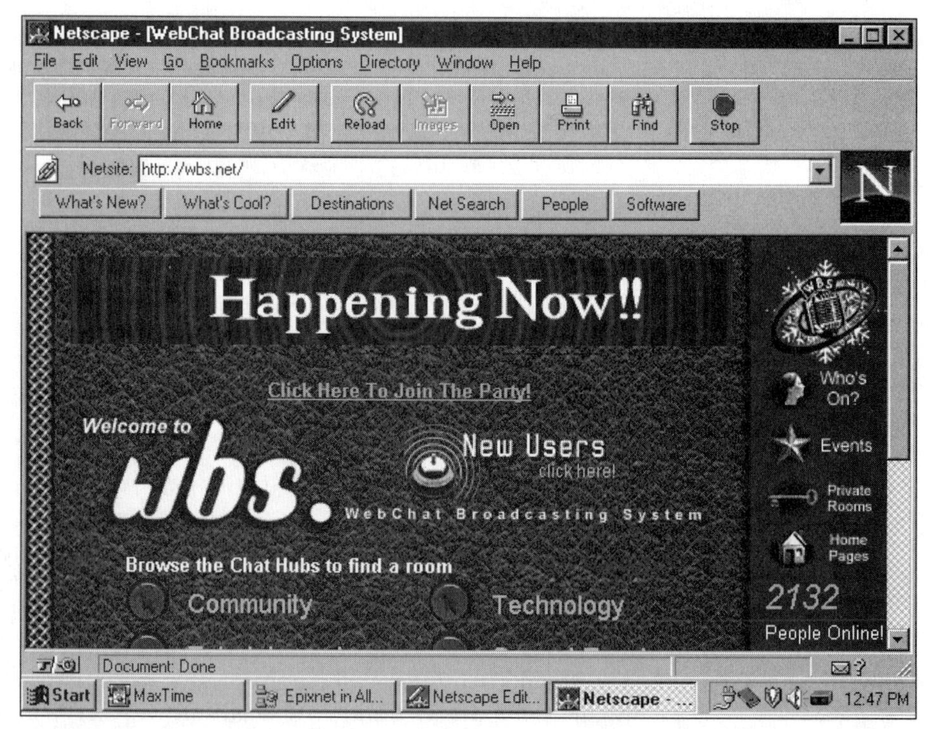

Figure 7-5.
Sample WebChat Station Page

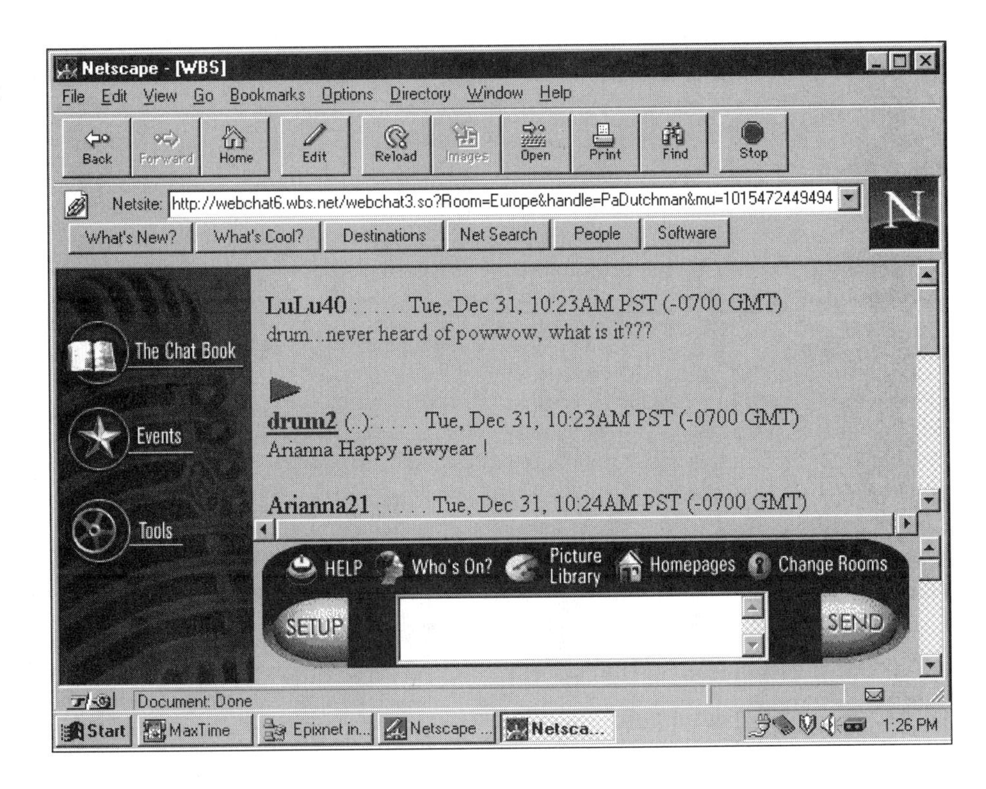

come page will appear. Click on the Chat Now button and a page will appear with the current participants identified by their handles. A graphic and message will also appear. (See Figure 7-5 for a sample page.)

6. Scroll down to the bottom of the main page to fill in the information. Use 10 as the refresh rate. (The *refresh rate* is the amount of time that will elapse before WebChat reloads the page.) Add a message in the white block at the bottom. Optionally, you may scroll down to add a URL for a graphic file and an e-mail address or the URL for your homepage. If you add your e-mail address or URL, participants who click on your hypertexted handle will be linked to them. (See Figure 7-6 for an example of information filled in on a page.)
7. Click on the Send button to add your data to the page. (The page should reload.)
8. Scroll down to find your entry. Figure 7-7 shows the information in Figure 7-6 displayed.

If you want to change the information or respond to someone else's message, scroll to the bottom of the page and click on the word Pause. Change the information, then click on the More Messages button.

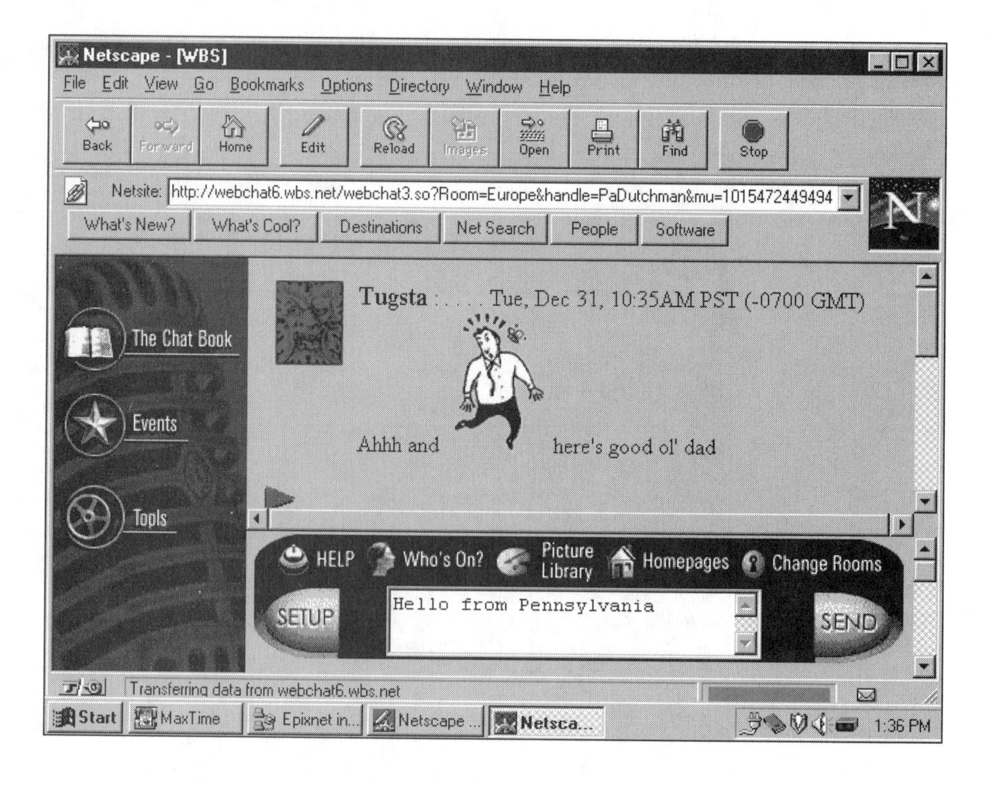

Figure 7-6. Example of WebChat Information on a Station Page

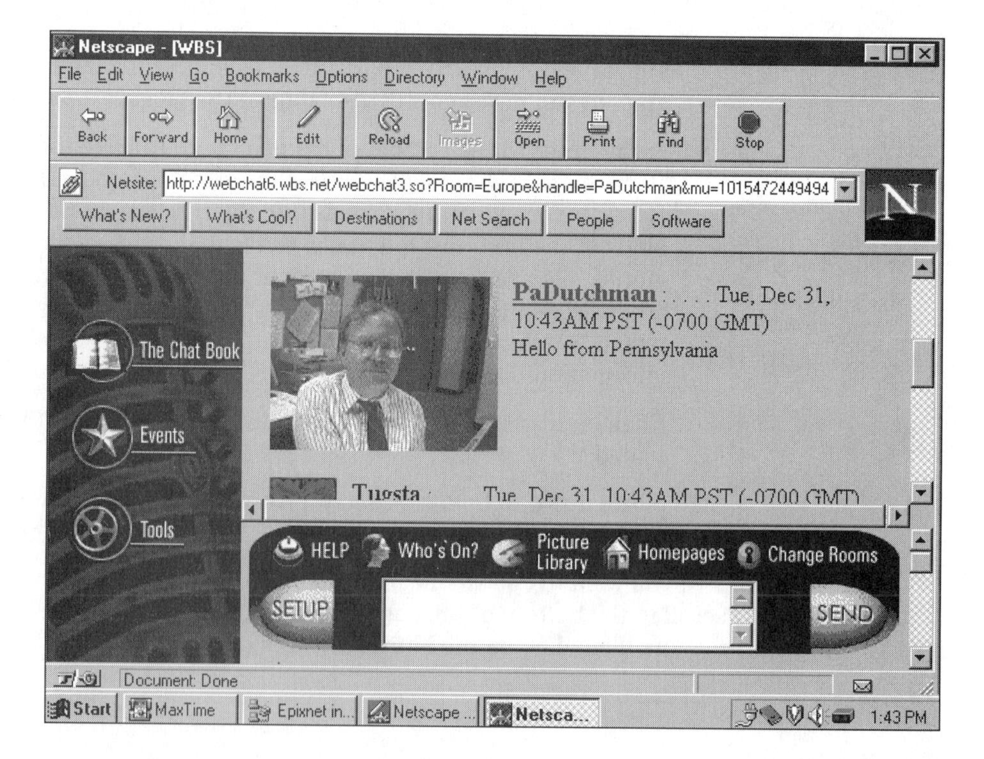

Figure 7-7. Displayed WebChat Station Page Information

The following are some tips for using the WebChat program.

- If you want to change to another hub, click on Change Rooms and your browser will go back to the main page. Click on a new hub name.
- Clicking on Events on the main WebChat page will provide you with a list of special interest discussion groups. These events may relate to relevant topics pertaining to computers, medicine, music, current events, and the like.

Telephony Computer Conferencing Programs

Telephony computer conferencing programs allow users to speak in real time to anyone anywhere in the world with a similar telephony phone connection. There has been controversy about these programs because users don't need to pay the normal long distance charges they would incur using the normal telephone network. As a result, the phone industry has challenged the legality of telephony conferencing programs. It is doubtful, however, that these programs will be prohibited from usage since the Internet is a public network. We will examine two telephony conferencing programs: Internet Phone and CoolTalk.

The Internet Phone Program

Internet Phone from VocalTec is a program that represents a type of computer conferencing in which you are able to speak, with your own voice, with Internet users in real-time, real-voice conversations. You can speak to anyone in any location with an Internet connection, a Windows-compatible audio device (sound card), and the Internet Phone program. A full duplex sound card is recommended, but a half duplex card will work if the two parties don't attempt to speak at the same time.

You can use Netscape or another browser program to download the Internet Phone program at http://vocaltec.com for a free evaluation copy that can be installed and used immediately. The evaluation copy allows you to use all of the program's features, but you are limited to 60 seconds of speech. You are permitted to use the program for 30 days before registering and paying a registration fee.

To use the program, connect to the Internet, plug in a microphone and speaker to your sound card, and run the Internet Phone software. Internet Phone can be used to meet new friends, get information personally, or make a direct business contact. A friendly graphic user interface and a smart Voice-Activation feature make conversations easy. VocalTec's sophisticated voice compression and transfer technology makes sure your voice gets across in a flash using only a fraction of the bandwidth.

Steps and Procedures for Using Internet Phone

Follow these steps to use Internet Phone.

STARTING THE PROGRAM AND CONNECTING TO A SERVER

To start the program and connect to an Internet Phone server, follow these steps.

1. Start the program. After the initial title screen, the main interface screen should appear (see Figure 7-8).
2. Select Connect To Server from the Phone menu or from the Tool bar. (The Connect To Server dialog box appears; see Figure 7-9).
3. The first time you try to connect to the Internet Phone network, the Internet Phone Servers dialog box automatically opens. This dialog box displays a list of countries and Internet Phone servers. It also includes a list of servers that are publicly accessible from any country. Choose your country, and from the list that opens, choose an Internet Phone server. Now choose OK, and the selected server is added to the dialog box.

Figure 7-8. The Initial Internet Phone Interface Screen

Figure 7-9.
The Internet Phone Connect To Server Dialog Box

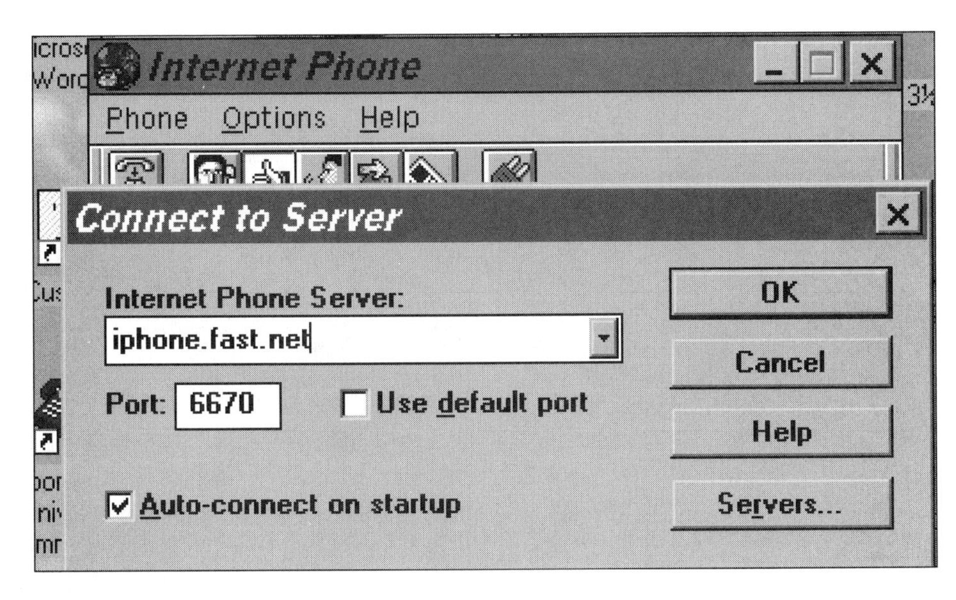

To access this dialog box again, choose the Servers button. If you want to enter the Internet address of an Internet Phone server manually, type it in. Internet Phone also keeps a list of the last five servers to which you connected, letting you choose one of them. You can connect to the Internet Phone network through any Internet Phone server. However, if you want to reduce response time and increase performance, you should try to connect to an Internet Phone server in your vicinity.

If the Internet Phone server to which you want to connect does not use the default port, uncheck the Use default port option and type in the correct port number. If you want to connect automatically to this Internet Phone server each time you run Internet Phone, check the Connect Automatically on Startup option.

4. Choose OK (see Figure 7-10). Internet Phone tries to connect to the Internet Phone server. During the connecting process, the IPS indicator blinks in yellow. When the connection is successful, the indicator turns green. If the Nick Name you entered in User Info is already used by another person, you will have to change it. Choose User Info from the Options menu and enter a new Nick Name. The first time you connect to the Internet Phone network, Internet Phone joins you to the general topic. The next time you connect to the network, Internet Phone automatically joins you to the topics under which you were listed when you last exited the Internet Phone network.

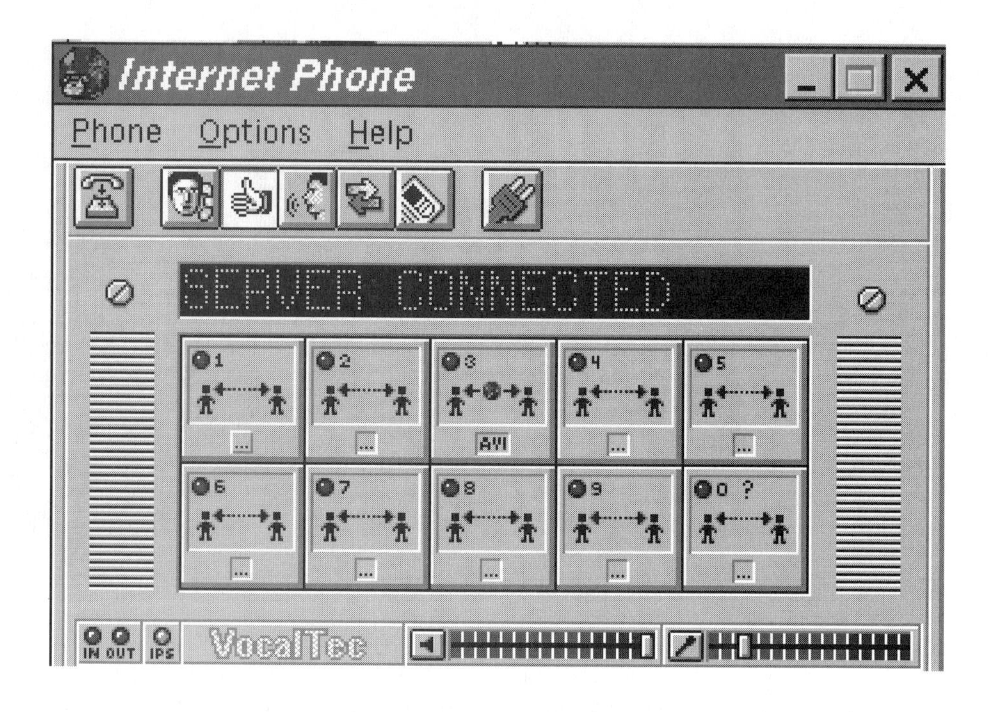

Figure 7-10. Internet Phone with Server Connection

When you disconnect from the Internet Phone server, Internet Phone saves to disk the names of the topics under which you were listed and will automatically list you under these topics the next time you connect to the network.

JOINING OR CREATING A TOPIC

To join or create a topic, follow these steps.

1. Select Call from the Phone menu, or choose the toolbar Call button. (The Call dialog box opens. The topics you are now listed under appear on the Joined Topics list on the right. See Figure 7-11.)
2. Choose the Join Topic button. (The Join Topic dialog box opens. This dialog box gives a global list of all the available Internet Phone topics. Of course, this list constantly changes as users create new topics or leave existing ones.)
3. Since updating the topic list can take some time, you must choose the Refresh button to get the updated list of topics. If you already know the name of the topic you wish to join, you can simply type its name in the Topics box without having to update the entire list.
   - To join one of the topics on the list, select it.
   - To create a new topic, type its name in the Topics box even without updating the list. Of course, if such a topic already exists, you will simply join it.

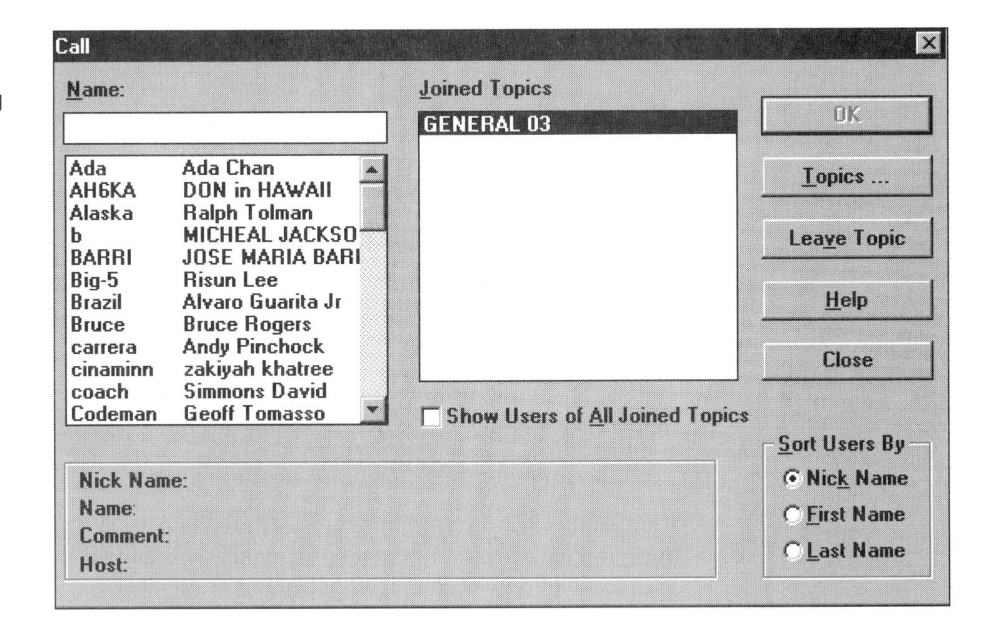

Figure 7-11. The Internet Phone Call Dialog Box

4. Choose the Join Topic button. The selected topic is added to the Joined Topics list on the other dialog box and the other users listed under it appear on the Users list. (To join another topic, repeat steps 3 and 4.)

5. To exit the dialog box, choose Close. (The Join Topic dialog box closes.)

You can now call one of the persons in the topics you have joined or wait for others to join and call you. When you leave a topic, the other users connected to it cannot contact you, unless they share other topics with you or have set your name under a Quick-Dial button.

LEAVING A TOPIC

To leave a topic, follow these steps.

1. Select Call from the Phone menu, or choose the toolbar Call button. (The Call dialog box opens. The topics you are now listed under appear on the Joined Topics list on the right.)

2. Select the topic you wish to leave from the Topics list.

3. Choose the Leave Topic button. (Your name is removed from the topic.)

If you don't want to be accessible to all the other Internet Phone users, you and other users can use private topics to access each other. Private topics are identical to listed topics but cannot be seen in the global topic list. Only users that know the exact private topic name are able to join it. Private topics can be used by members of worldwide organizations and groups to access each other without interference.

Of course, for a private topic to be of any use, it is preferable that you remove your name from any listed topics. It is especially important to leave the General topic, or many other Internet Phone users will be calling you constantly.

JOINING A PRIVATE TOPIC

To create or join a private topic, follow these steps.

1. Select Call from the Phone menu, or choose the toolbar Call button. (The Call dialog box opens. The topics you are now listed under appear on the Joined Topics list on the right.)
2. Choose the Join Topic button. (The Join Topic dialog box opens. This dialog box gives a global list of all the available Internet Phone topics.)
3. Type the private topic's name in the Topics box.
4. Choose the Join Private button. (If there is no such topic, it will be created. If a listed topic of that name exists, you will be informed. The private topic is now added to your Joined Topics list.)
5. To exit the dialog box, choose Close. (The Join Topic dialog box closes.)
6. You can now call one of the persons listed under the private topic or wait for others to join it and call you.

You can use Internet Phone to call and talk with other users who are connected to the Internet Phone network and have Internet Phone running. You can call a person with the Call command or with a Quick-Dial button set for him or her. (If you want to conduct a full duplex conversation with the other person, make sure the Full Duplex option is on before you make the call.)

CALLING A PERSON

To call a person, follow these steps.

1. Select Call from the Phone menu or choose the toolbar Call button. The Call dialog box opens. The Joined Topics list displays the topics under which your name is now listed. The Users list displays the other persons listed under the topic selected in the Joined Topics list. To see the names of all the users listed under all the joined topics, check the Show Users of all Joined Topics option.
2. Locate the desired person in one of the joined topics. If you cannot locate the person, you can try looking for him or her under other topics.
3. Select a person from the Users list.
4. Choose OK. (A message appears on the message line, accompanied by a dialing sound, indicating that the call is in progress. See Figure 7-12.)

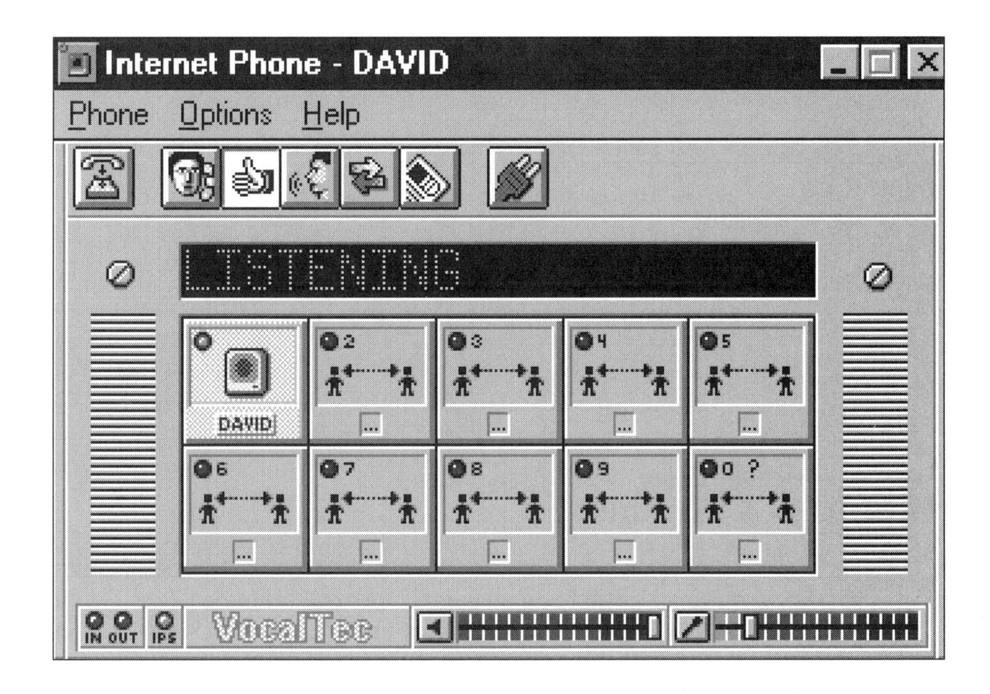

Figure 7-12.
Internet Phone Calling a Person

One of the Quick-Dial buttons shows a hand knocking on a door. This button will be used for interaction with the person. If one of the buttons already holds the person's name, it will be used. If all the buttons already hold other names, the Wild Card button will be used.

On the other person's Internet Phone, a Quick-Dial button representing you starts ringing. He or she can answer your call by clicking on this button. If your call is answered, you will be able to talk with the other person; if the person is already engaged in another conversation, the message line will report BUSY. When contact with the other person's system is established, conversation starts.

If you want a much faster way to call, set a Quick-Dial button for that person. If one of the Quick-Dial buttons is empty, it is automatically set to hold the person's name. Whenever you want to call the person, simply click on the Quick-Dial button. This procedure is not only much faster, but also lets you contact the person even if you are both listed under different topics.

When another person calls you, the Quick-Dial button representing the caller changes its appearance to that of a ringing phone, accompanied by a ringing sound. The message line shows the caller's name. If none of the Quick-Dial buttons is set to represent this caller, the Wild Card button is used to interact with the caller.

If the Bring to Top on Call preferences option is on, the Internet Phone window will pop up whenever a person calls you.

RECEIVING A CALL

To receive a call, do one of the following:

- To accept the call, choose the caller's Quick-Dial button.
- To reject the call, choose Disconnect from the menu or toolbar.

If you accept the call, conversation with the other person starts. It is possible to make access to your system immediate by turning on the automatic confirmation mode. If another person tries to call you while you are speaking with someone, the message line shows the caller's name, and the caller gets a BUSY message. Conversation starts when contact is established between two persons. The message CONNECTED! appears on the message line.

The CoolTalk Program

CoolTalk is a program similar to Internet Phone. The Netscape Corporation provides CoolTalk as a separate program bundled and installed automatically for you with Netscape 3.0 and above. CoolTalk can be downloaded as a component of Netscape at http://home.netscape.com/comprod/mirror/index.html. CoolTalk is installed as a helper application for Netscape that is executed from within Netscape when you link to another CoolTalk user. Figure 7-13 shows the CoolTalk console.

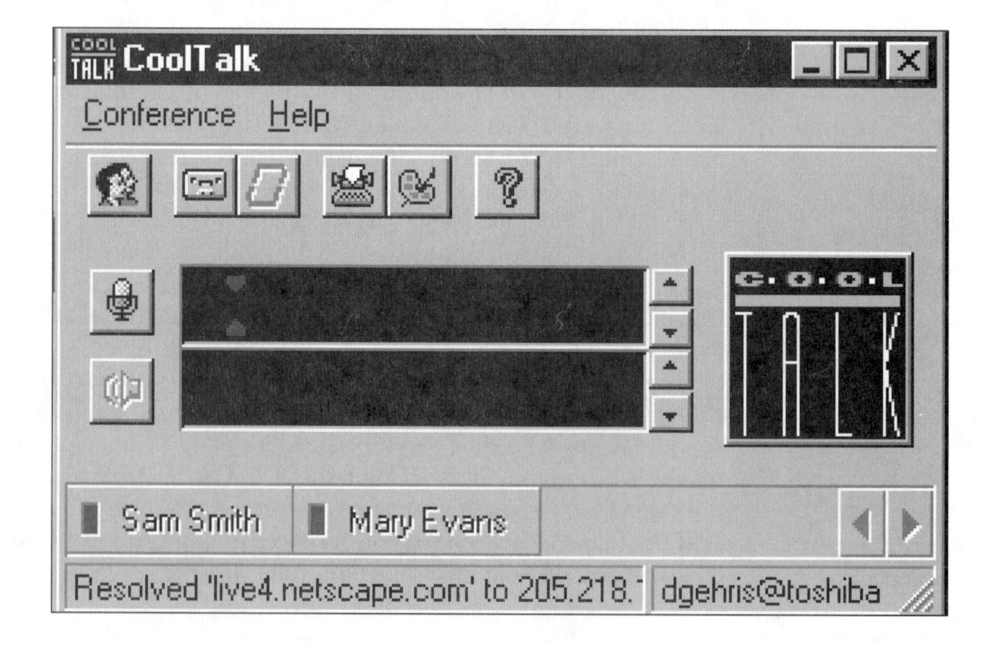

Figure 7-13. The CoolTalk Console

Like Internet Phone, CoolTalk serves as an Internet telephone, complete with audio conferencing–chat capabilities that allow you to talk with friends and associates across the Internet with full duplex sound so that you can speak and be heard simultaneously. CoolTalk includes a speed dialer, caller ID functionality, call screening, and mute buttons.

SETTING UP THE BUSINESS CARD

The first step is to set up your CoolTalk business card. You will be prompted to set it up when you access CoolTalk. Items can be changed by choosing Options on the Conference menu and then clicking on the Business Card tab. On the business card you specify information that you want other CoolTalk users to view. There is an option to link a picture to the card. This picture will appear on the other participant's CoolTalk console and you will see his or her picture on your console. Figure 7-14 shows a sample business card.

CoolTalk has some features that make it different from other Internet telephone programs, including a phonebook, shared whiteboard, and answering machine.

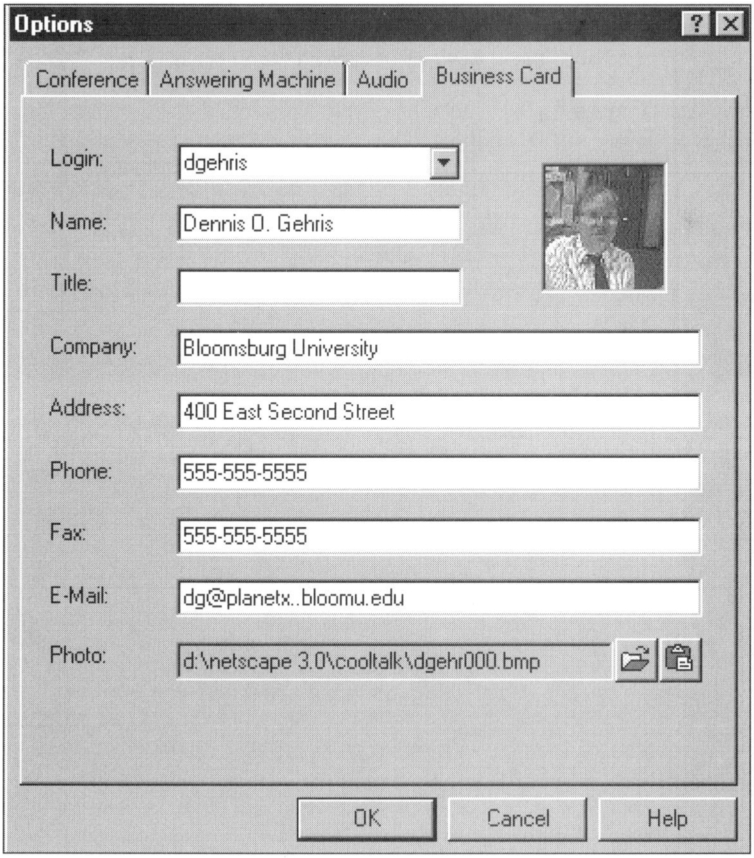

Figure 7-14. Sample CoolTalk Business Card

PHONEBOOK

One of the unique features of CoolTalk is a Web-based phonebook that makes it easy to locate other CoolTalk users. The URL for the phonebook is http://live.netscape.com. You can list users by location or alphabetically. You simply use Netscape to link to the phonebook page, find the user that you'd like to call, and click on his/her name. The CoolTalk console will be displayed and the built-in phone dialer will call the person. If the user you've chosen is available, he or she will respond and the conference can begin. If another user contacts you while you have CoolTalk running, a window will pop up asking you to accept or reject the invitation to enter into a discussion. Figure 7-15 shows a portion of the CoolTalk phonebook.

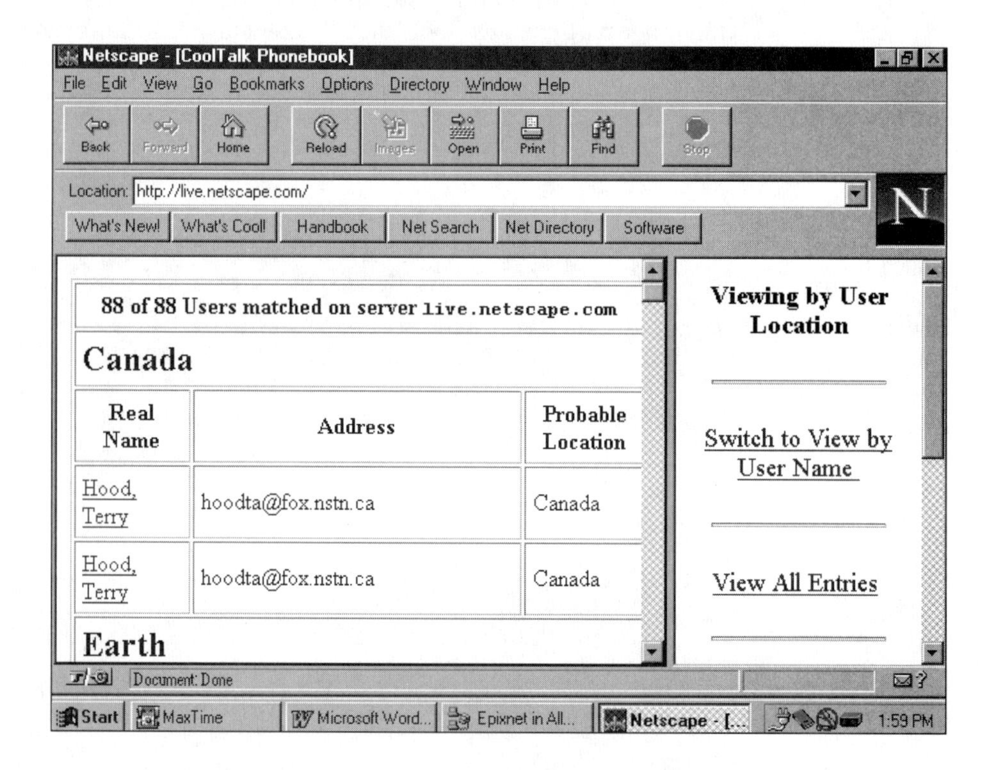

Figure 7-15. The CoolTalk Phonebook

WHITEBOARD

The shared whiteboard can be used for textual and graphical data conferencing. While you talk with a colleague via CoolTalk, both persons can view and edit graphics in real time with a full range of drawing, markup, and zoom tools. It is also possible to import, save, and print TIFF, GIF, JPEG, BMP, TARGA, EPS, SGI, and RASTER files. Figure 7-16 shows the CoolTalk whiteboard with sample data.

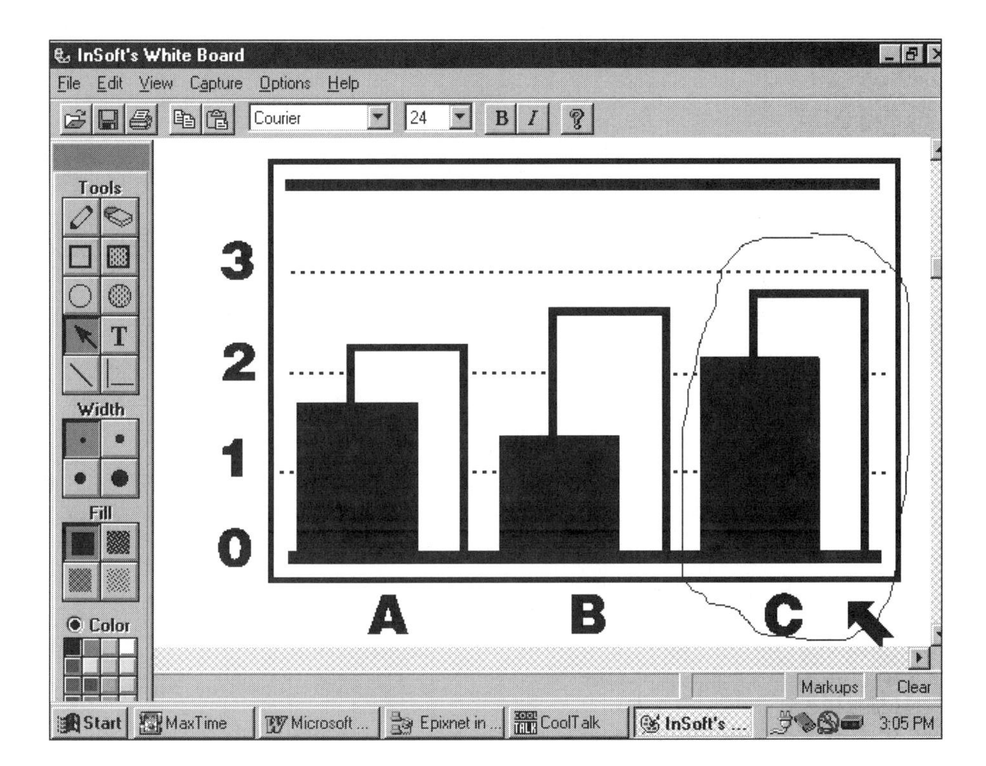

Figure 7-16. The CoolTalk Whiteboard with Sample Data

ANSWERING MACHINE

CoolTalk's answering machine feature enables users to send you messages if you are away from your computer and they attempt to conference with you. In much the same way that a real answering machine operates, the CoolTalk answering machine connects after a few rings and plays an outgoing message. It then records a message from the calling party and saves it to a file on disk. When you return to your computer, you can play back the recorded message. The answering machine is found by clicking on the Answering Machine tab on the Options item on the Conference menu. Figure 7-17 shows the Cooltalk answering machine.

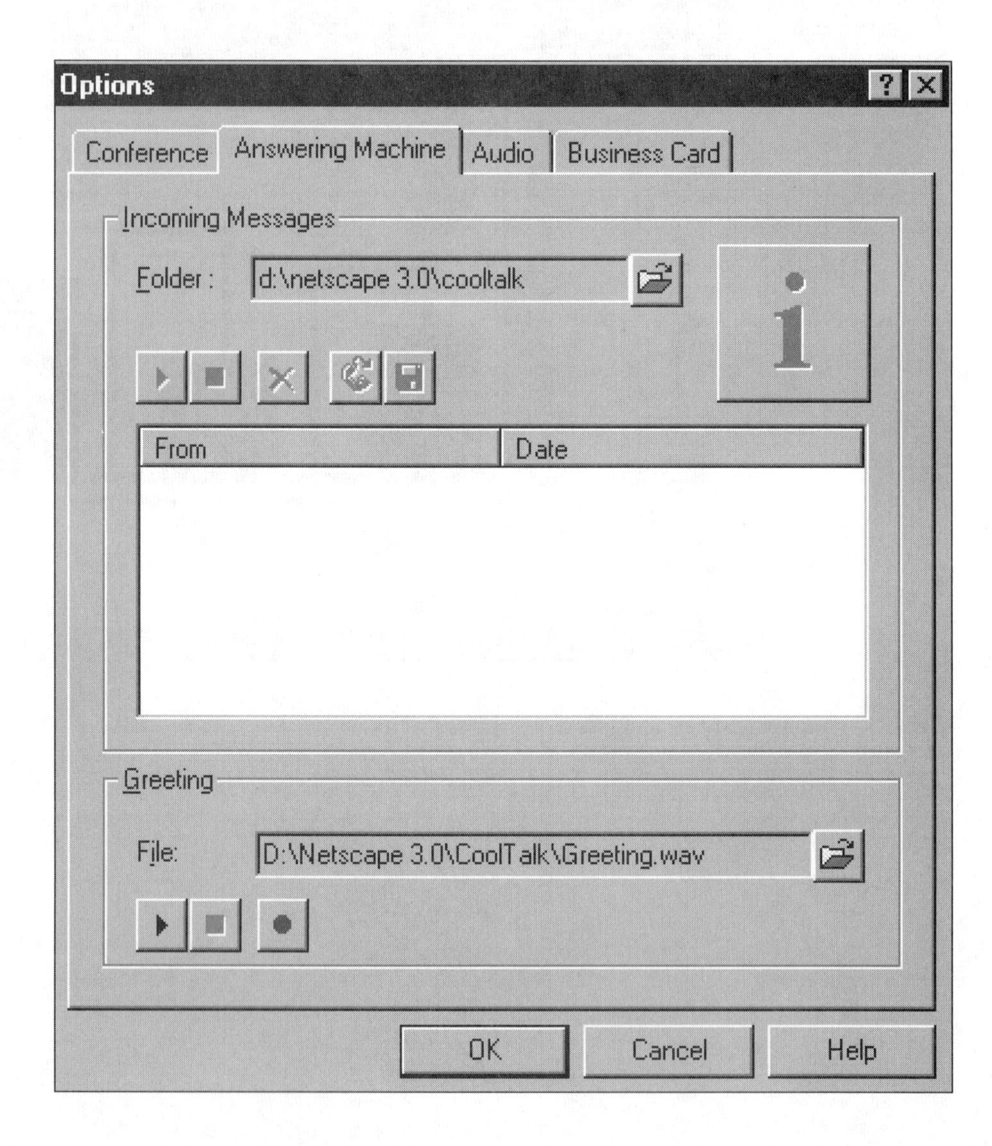

Figure 7-17. The CoolTalk Answering Machine

Questions for Review

1. What is computer conferencing? How is it different from electronic mail?
2. What are the advantages of computer conferencing versus face-to-face conferences? What are the disadvantages?
3. What are the two main types of computer conferencing application programs? How do they differ?
4. What are the two main types of UNIX-based computer conferencing programs? Explain how each operates.
5. Explain the function of the following IRC commands: /list, /join. Investigate the other IRC commands and make a list of them.
6. What is the purpose of Web pages sponsored by IRC channels?
7. What are the characteristics of Web-based computer conferencing programs? How do they differ from the UNIX-based programs?
8. How can individuals use Web-based computer conferencing programs? Provide examples.
9. How can businesses use Web-based computer conferencing programs? Provide examples.
10. What are the characteristics of Netscape Chat? How does it operate?
11. What are the characteristics of WebChat? How does it operate?
12. What is a telephony computer conferencing program? Why has it created controversy?
13. What are the characteristics of Internet Phone? How does it operate?
14. What are the characteristics of CoolTalk? How does it operate? How is it different from Internet Phone?

Exercises

1. Download and install the Netscape Chat program.
2. Use the Netscape Chat program to connect to a chat server. Join a chat channel that interests you, participate in a discussion, and try sending a URL containing your school's Web homepage. Write a paragraph that describes the discussion that took place. What were the benefits of communicating with Netscape Chat in this manner?

3. Use a scanner, if available, to scan a photograph of yourself or a building on your campus and create a graphic file (see Chapter 4). Have the graphic file downloaded to a server on campus, or locate a URL of an existing graphic file on a server on the Internet and note the URL for use with WebChat.
4. Use Netscape or another browser to access WebChat. Find an interesting station and participate in a discussion. Specify the URL for a graphic file as a part of the message. Write a paragraph that describes the discussion that took place. What were the benefits of communicating with WebChat in this manner?
5. Use a search engine to conduct a search using the word Chat to locate channels or stations. Make a list of these topics along with a description of each.
6. Use Netscape or another browser to link to the WebChat Events Page (http://wbs.net) and click on Events. Make a list of events that are occurring during the current month. Is there a listing relating to the Internet? If so, what is the current topic of discussion?
7. Download and install the Internet Phone or CoolTalk programs.
8. Connect to an Internet Phone server and join a topic of discussion, or use CoolTalk to contact a person listed in the phonebook. Use the program to contact and converse with a specific person who is also using the Internet Phone program.

Case

Delicious Candy Company manufactures specialty candies and markets them to major department stores, gift and card shops, and retail candy stores. It employs 120 salespeople who are each assigned a territory within six regions in the United States. The company will be introducing a new line of chocolate liqueurs to complement their candy line. Write a proposal that describes how this firm might use Netscape Chat, WebChat, Internet Phone, or some other multimedia computer conferencing program for training their sales staff on the details associated with selling this new product. Make sure to include the benefits of using this type of conferencing over face-to-face conferencing.

CHAPTER

8

Videoconferencing on the Internet

This chapter will explore videoconferencing as an additional multimedia application on the Internet. It will discuss videoconferencing applications and how to conduct a successful videoconference, explain the hardware requirements and procedures and the steps for using CU-SeeMe videoconferencing from Cornell University, and compare the various Web-based conferencing application programs reviewed in this textbook.

What You Will Learn

- How videoconferencing is defined
- Sample applications for videoconferencing
- How to conduct a successful videoconference
- Steps and procedures for using CU-SeeMe videoconferencing
- How to obtain more information about CU-SeeMe videoconferencing
- How conferencing application programs compare with one another

Videoconferencing Defined

A sophisticated type of teleconferencing that has recently attracted much attention for use in business and in education, **videoconferencing** is a full-motion, two-way, video/audio system that permits two or more people in different locations to communicate with each other. This system contrasts with what is known as business television or point-to-multipoint videoconferencing, in which all participating sites receive video images but only one site is able to transmit them. Two-way videoconferencing is often used for large groups and by colleges and universities that offer video courses.

Companies have, however, experienced several barriers to using videoconferencing. Formerly, the costs of equipment as well as transmission costs for running a videoconferencing system were very high; consequently, only a few very large organizations were able to afford to install videoconferencing systems. Today, however, the costs have greatly decreased so that smaller organizations have been able to enjoy the benefits of videoconferencing. An ongoing problem, however, is the lack of a national broadband network that will make it possible to send multiple applications over the telephone lines simultaneously. Until a broadband network is operational, use of the integrated services digital network (ISDN) or digital simultaneous voice data (DSVD) technology, as discussed in Chapter 6, is desirable.

There are several possible facility configurations for videoconferencing. One, used by larger organizations, requires built-in or permanent videoconference rooms that contain cameras and other video equipment but otherwise resemble a regular conference room. Another is the mobile or portable conference system contained in a unit that can be moved from one room to another. The third is the desktop system, in which a personal computer is combined with a camera, microphone, and network connection. Special computer software is also needed to make the desktop videoconferencing system complete.

Videoconferencing Applications

Videoconferencing can be used in business, education, and personal communication in many ways. The discussion here only touches on a few possible applications for this exciting technology.

BUSINESS

Larger businesses use videoconferencing because it is less expensive than traditional methods of bringing people together. After allocating funds for purchasing equipment and software, selecting transmission methods, and hiring and/or training personnel to set up the system, the cost savings over the long run can be substantial. Moving employees from place to place is often very expensive, and work time is lost in the process. Businesses typically use videoconferencing for employee training, committee and work group work, or to introduce a new product, service, or procedure.

EDUCATION

The term **distance learning** is often used in education to denote the use of videoconferencing to link several classrooms containing students or a scattered group of individual students together with one common instructor or lecturer. Some school districts hope to use distance learning to broaden course offerings. If, for example, enrollment does not warrant employing French teachers for each of three high schools and junior high schools in a district, one teacher can be hired to link to each of the schools via videoconferencing.

Some schools have also experimented with distance learning as a supplement to home instruction. In this way, students who are unable to travel to school because of health reasons, distance, or parental preference can hear and participate in classes at home.

Colleges and universities now use technology to broadcast what are commonly called *telecourses*. Telecourses allow a student to view a series of videotaped lectures over a local television cable system or broadcast. After completing other requirements, such as taking tests and submitting research papers, the student receives the usual course credit. Some of the larger colleges have been experimenting with supplementing or replacing telecourses with live broadcasts of lectures and seminars so that the telecourses become a type of distance learning component.

Bill Gates of the Microsoft Corporation envisions schools installing what he calls digital whiteboards in classrooms. The whiteboards would project images from the information highway and could be used for videoconferencing, making it unnecessary to install large video screens or other projection devices.

PERSONAL USE

There are also many potential personal-use applications for videoconferencing. It could be used to bring together friends who because of distance aren't able to get together as frequently as they would like. Children or adults could use it to play video games or to research a topic by videoconferencing with an individual who possesses expertise in the subject area. In the future, videoconferencing is likely to take the form of interactive television in which individuals are able to express opinions relating to an issue or question being discussed by pressing a button on a hand-held device.

Conducting a Successful Videoconference

Yvonne Marie Andres (e-mail andresyv@cerf.net), director of the Global Schoolhouse, offers a number of suggestions on conducting a successful videoconference. Her tips focus on the use of lighting, cue cards/eye contact, audio, photos and props, interactivity, appearance and attention, moderators, video clips, and rehearsals.

LIGHTING

Videoconference organizers should not rely on overhead lights, especially fluorescent lights, and should make sure that a window is not directly behind the speaker; otherwise, he or she will appear as a dark shadow. You should try not to squeeze more than two or three people into your video window because it is important to be able to see speakers' facial expressions as they talk. You also should avoid using a portable light that you shine on the speaker in a "spotlight" fashion.

CUE CARDS/EYE CONTACT

A speaker's notes should be printed on poster boards in a very large font to make cue cards. This will prevent the audience from viewing the speaker's forehead instead of his or her eyes. The speaker should look directly into the camera while speaking. The person holding the cue cards should stand behind the camera so that the speaker appears to be looking into the camera while accessing notes.

AUDIO

To make speakers' voices easier to hear given the poor audio transmission over the Internet, you should pause frequently in your video transmission and have speakers talk at a slow rate, speaking loudly and distinctly. Speakers should vary the tone of their voices, using inflection while they speak.

PHOTOS AND PROPS

Speakers should use photos and props to enhance their presentations, thinking of themselves as storytellers. A simple way to accomplish this is to hold the photo in front of the camera and pause the picture. After resuming, the speaker tells the story that goes with the picture. This approach is much more effective than watching a so-called "talking head."

INTERACTIVITY

You are advised to stop every 10 minutes (or less) to allow for interaction. The main reason for using "live" video is so that viewers can interact with the speakers.

APPEARANCE AND ATTENTION

Because it is distracting to see participants fidgeting, yawning, or talking among themselves, this behavior should be avoided. If participants feel they must engage in these actions, the video transmission should be temporarily suspended. Some clothing shows up better than other clothing on video transmissions. Videoconference organizers should experiment before the event to see what fabric colors and patterns look best.

MODERATOR

Use of a moderator for a videoconference is highly recommended, and that person should play a major part in the conference. The moderator should make the welcoming comments, cue each participant when it is his or her time to speak, keep the pace of the conference moving along, and end the conference as planned.

VIDEO CLIPS

Very short (less than 2-minute) video clips can be used to enhance the videoconference. Keep in mind that the video clip will appear to be progressing much more slowly to the viewers than it may to you.

REHEARSALS

Rehearsals are a very good idea, especially when dealing with inexperienced presenters. Participants should practice in front of the camera and critique one another several times before an event.

Introduction to CU-SeeMe Videoconferencing

A major free two-way videoconferencing software being used over the Internet is **CU-SeeMe** (pronounced "See you, See me"), developed and distributed by Cornell University. The first network-capable version of CU-SeeMe, called WatchTim, was created in August, 1992. CU-SeeMe developers at Cornell were then asked to assist in setting up an Internet videoconference sponsored by the National Science Foundation in 1993. The developers have also had a relationship with the Global Schoolhouse, a project that links schools with business and government throughout the United States. Recently, White Pine Software has teamed up with Cornell to further develop the CU-SeeMe software. White Pine Software is marketing an enhanced color version of CU-SeeMe. For more information on this version, visit their homepage at http://goliath.wpine.com /cu-seeme.html. See page 181 for more information on Enhanced CU-SeeMe.

CU-SeeMe software is currently available for both Macintosh and Windows (PC-compatible) computers. Users can connect to one another by point-to-point so that the connection operates somewhat like a video phone. To use CU-SeeMe by point-to-point, two individuals must be connected to the Internet and use CU-SeeMe software simultaneously. Each will see a video window of the other, and they should be able to speak back and forth with a microphone. One user must know the numeric IP address of the other, however. When an Internet service provider provides IP addresses to users dynamically, so that a different address is assigned each time one connects, there must be a means for determining the IP address.

CU-SeeMe for group conferencing requires the use of a *reflector*, a computer using the UNIX operating system that runs special software. In this way, several video windows appear on the screen, one for each person connected to the reflector. The reflector redistributes video and audio streams to each user who connects to it. Because of the amount of bandwidth required for a group conference, a regular phone connection using a 28.8 Kps modem may not yield good results. A direct Internet connection is much better for group conferencing with CU-SeeMe.

CU-SeeMe can also be used for one-way broadcast sessions. For example, the National Aeronautics and Space Administration broadcasts their NASA TV carried on several CU-SeeMe reflectors. The broadcasts show ongoing NASA operations, including press conferences, launches, landings, and the like. Some of these broadcasts are broadcast live, others are videotaped. Figure 8-1 shows a video window from a NASA TV CU-SeeMe broadcast.

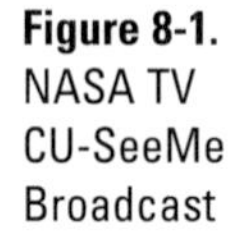

Figure 8-1. NASA TV CU-SeeMe Broadcast

Special Hardware Requirements for CU-SeeMe

The special hardware required for CU-SeeMe falls into two categories: video and audio. This is in addition to the basic hardware discussed in Chapter 1. Video equipment is needed to send video images to others who are participating in a videoconference. Although CU-SeeMe will work without video equipment, its effectiveness will be greatly diminished. The audio equipment may be necessary to hear others talk to one another and for you to participate in the conversations.

VIDEO EQUIPMENT

The most desirable piece of equipment is some type of video camera. Video cameras can be classified into two categories: dedicated desktop cameras and general-purpose analog cameras. Dedicated cameras are those designed specifically for desktop applications and are generally more expensive than general-purpose cameras. To use a general-purpose camera and some dedicated cameras with a PC-compatible computer, you will also need a video capture card. Not all video capture cards work with CU-SeeMe. Check the compat.txt file that is a part of the CU-SeeMe documentation for a list of compatible cards.

One desktop camera that works with CU-SeeMe is the Quickcam by Connectix, the best-known, inexpensive grayscale and color image cameras currently available for both the Macintosh and Windows (PC-compatible) computers. Quickcam is a golf-ball size camera that does not require a video capture card. On the Macintosh, it works with QuickTime software and connects to the computer via a serial cable to your printer or modem port. On the PC-compatible computer, it uses Video for Windows software and connects via the parallel port and keyboard connector.

The grayscale camera provides images up to 320 by 240 pixels in 16 shades of gray and operates at up to 15 frames per second. It has an onboard microphone, a field of view of about 65 degrees, and a fixed focus. The color camera has a 5.7mm lens with a 48-degree field of view that provides resolutions up to 640 x 480 pixels in lifelike 24-bit color. Although the microphone is functional, it competes for bandwidth with the video signal. For that reason, you may want to consider a separate microphone, as discussed in the next section.

For more information on the Quickcam camera, link to http://www.connectix.com or to http://baby.indstate.edu/mstattler/sci-tech/comp/hardware/quickcam.html. Figure 8-2 shows the Quickcam.

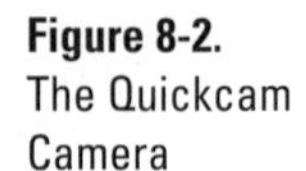

Figure 8-2.
The Quickcam Camera

AUDIO EQUIPMENT

The following audio options are available to Macintosh users: the external microphone or the PlainTalk microphone appropriate to the machine; the built-in microphone; third-party microphones; and the general-purpose desktop video camera's microphone. Using an add-on microphone with CU-SeeMe will work on the Macintosh if the computer has a built-in audio input jack.

Audio support for CU-SeeMe for Windows is a recent addition to the software. Any PC-compatible computer with built-in sound capabilities or an added sound card should work. However, the card needs to be full duplex, which means that it must accept input and output audio at the same time. Most SoundBlaster cards, the most popular sound card for PCs, are half duplex at the writing of this book. Although a half duplex card will work, the audio signals will be interrupted by other signals being transmitted at the same time. Some SoundBlaster products now feature full duplex support. An add-on microphone will connect to a jack on the sound card or somewhere on the computer if the sound is built in.

Procedures and Steps for Using CU-SeeMe

This section will explain how to use CU-SeeMe videoconferencing software on either a Macintosh or an IBM or IBM PC–compatible computer.

DOWNLOADING AND INSTALLING CU-SEEME

CU-SeeMe for Windows (PC compatible) can be downloaded through Netscape at http://cu-seeme.cornell.edu/PC.CU-SeeMeCurrent.html. Another location is ftp://cu-seeme.cornell.edu/pub/cu-seeme. Open the PC.CU-SeeMeCurrent subdirectory with the latest date. Click on the Cuseeme.zip link. After downloading this file, you will need to use an unzip utility to unzip the file. The following files will result, some of which can also be downloaded separately. Place them all in a subdirectory of your hard disk drive (i.e., name it cuseeme).

- cuseeme.exe: the main program file
- changes.txt: what's changed in this release
- readme.txt: general CU-SeeMe information
- cuseeme.txt: the user's guide
- faq.txt: frequently asked questions
- compat.txt: list of compatible video capture cards and WinSock stacks, etc.
- msvideo.dll and ctl3d.dll: video driver files

Look in your windows directory (C:\WINDOWS) and in the windows system subdirectory (C:\WINDOWS\SYSTEM). Determine if you have copies of the msvideo.dll and ctl3d.dll files. If you do not have copies of these files, copy them from the CU-SeeMe subdirectory into your windows directory. If you already have copies, erase the files from the CU-SeeMe subdirectory.

Your PC-compatible computer will also need a hostname file called hosts, and the program may not work without it using Windows. To set it up, you must know the IP address that has been assigned to your computer. The hosts file is placed in the WinSock subdirectory. The format for the hosts file is <your IP address> <name for your PC>. For example, if your IP address is 192.23.43.99 and the name of your PC is Room_384, the entry in the hosts file would be: 192.23.43.99 Room_384. If you don't know your IP address, you may want to contact your network administrator. If your host name is defined in a domain name server (DNS) accessible to your PC, you won't need a hosts file. Be careful that only one hosts file exists on your hard drive. If more than one copy exists, verify which one is correct and delete the others. If a correct hosts file is not present on your hard drive, you may get a GetHostByName() Error message when you attempt to use CU-SeeMe.

You may also need to change your monitor's settings under Windows. Choose the setting that your digitizer card requires.

For the Macintosh, use http://cu-seeme.cornell.edu/get_cuseeme.html. Click on either CU-SeeMe for Macs or CU-SeeMe for Powermacs to download the necessary file. The Read Me First and Basic Mac Readme links contain current information about CU-SeeMe. You can also use the ftp://cu-seeme.cornell.edu/pub/cu-seeme URL. Open the Mac.CU-SeeMe subdirectory with the latest date. The file that you download will need to be decompressed by using the Aladdin's StuffIt Expander program. This freeware program is available on the Info-Mac and UMich software archives. You may make adjustments so that the display mode on your monitor shows at least 16 grays or 16 colors (4-bit video).

CONFIGURING CU-SEEME

Run CU-SeeMe by using the normal procedure for your computer. For PC-compatible computers, you will need to set up an icon. For Windows 95, click on Settings/Taskbar on the Start button. Then choose Start Menu Programs and click on the Add button. Enter the path for the cuseeme.exe file (i.e., c:\cuseeme\cuseeme.exe) and then choose the folder where you'd like the program icon to appear.

The first thing you'll need to do when you launch CU-SeeMe is to make changes in the Preferences dialog box. This box may appear automatically the first time that you launch the program. If it doesn't appear, choose Preferences from the Edit menu. Figure 8-3 shows the Preferences dialog box for the Windows version of CU-SeeMe.

Let's look more closely at each part of the Preferences dialog box.

Your Name	This is the name you want to appear at the top of your video window on other people's displays (i.e., Jake Smith@MSU). Change the name that appears and change it to your name.
Show Splash Screen at Startup	Displays the CU-SeeMe Splash screen in Receive Video windows until actual video data arrive.
Save Video Window Position	Saves the positions of the video windows currently on the screen.
Auto-Tile Video Windows	Automatically places video windows in order on the screen as they are opened.
Open New Video Windows	Automatically opens new video windows whenever a new sender joins the conference.
Max Video Windows	Shows the maximum number of video windows that can be opened at one time. This feature is most useful when you have selected Open New Video Windows (above) so that not too many windows are automatically opened at once. The minimum number of video windows that can be open is two and the maximum is eight.
Max kbits/sec	Shows the maximum number of kilobits per second transmitted by your workstation. This number is the total amount of data including video, audio, control info, and packet headers. Actual transmission rate is between Max kbits and Min kbits, depending on packet loss.
Min kbits/sec	Shows the minimum number of kilobits per second that your workstation will be limited to because of packet loss.

Next, check the video format (File/Video Setup/Video Format). CU-SeeMe needs to capture video using dimensions of 160 x 120 and an image format called *8 bit palletized*. Figure 8-4 shows the Image Size and Quality dialog box.

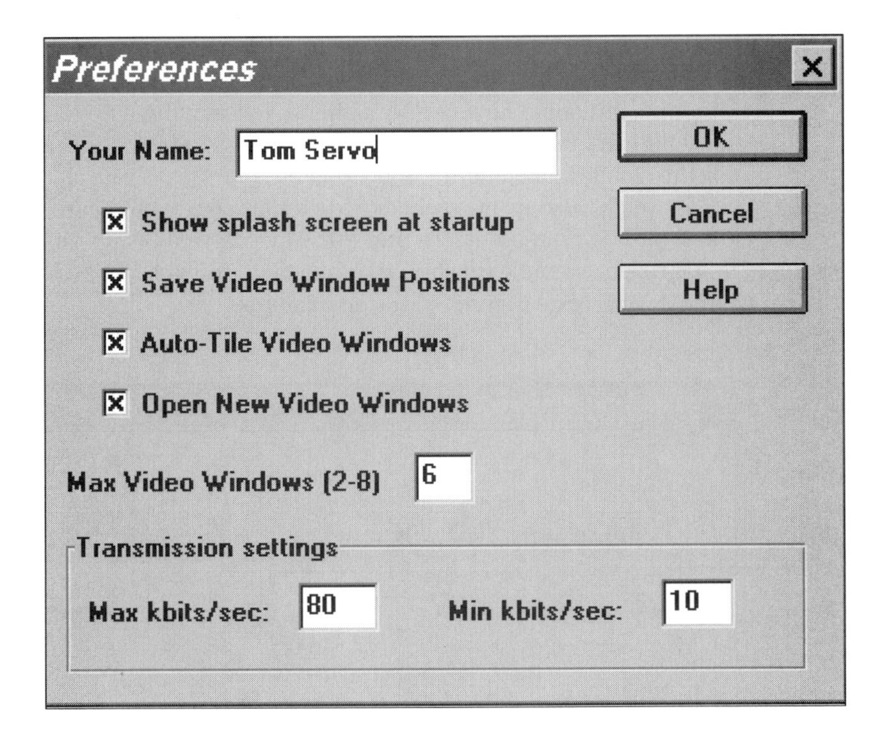

Figure 8-3. The CU-SeeMe Preferences Dialog Box

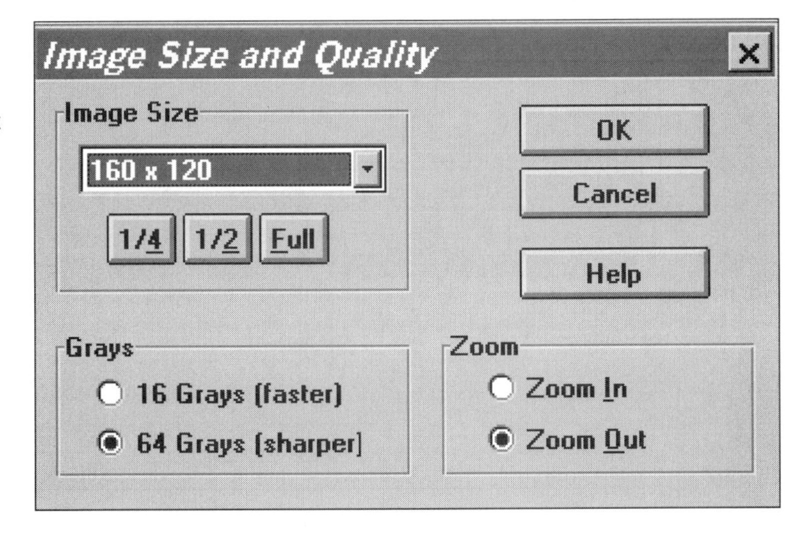

Figure 8-4. The CU-SeeMe Image Size and Quality Dialog Box

You may also want to check the video source (File/Video Setup/Video Source) to see that the correct video input is selected. Most users have NTSC cameras, but some have a PAL camera. You may be able to adjust the settings for your camera from this dialog box. Figure 8-5 shows the camera adjustments for the QuickCam camera.

You might also want to check to see that your sound card is set up properly. Click on File/Sound devices and the Wave Device Selection dialog box should appear. Your sound card should be specified for as both the recording and playback devices. See Figure 8-6 for an example.

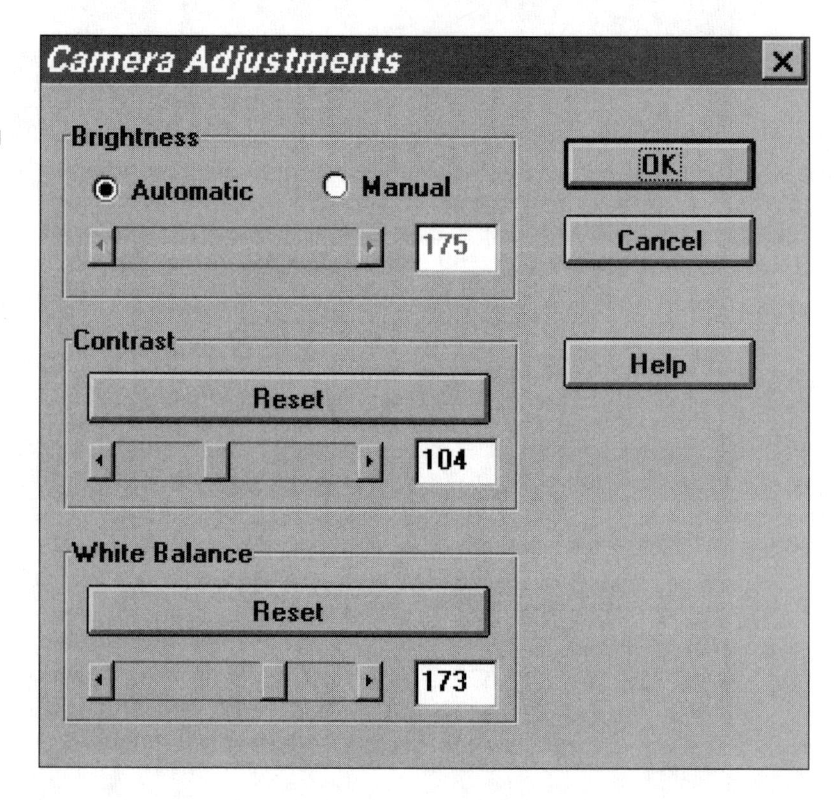

Figure 8-5. The CU-SeeMe Camera Adjustments Dialog Box for the Quickcam Camera

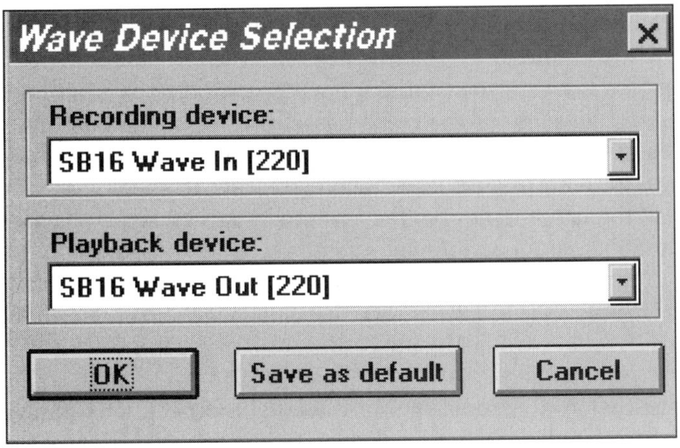

Figure 8-6. The CU-SeeMe Wave Device Selection Dialog Box

Wave Device Selection

Recording device:
SB16 Wave In [220]

Playback device:
SB16 Wave Out [220]

OK | Save as default | Cancel

STARTING CU-SEEME

If all of the settings are correct and you are connected to the Internet, you are ready to start CU-SeeMe. Initially, you should see the video window for your video camera. Depending on where you point the camera, you may see a video of yourself or another person or object in the room. Figure 8-7 is an example of what you should expect to see when you start CU-SeeMe. Your video window would, of course, replace the one that is shown.

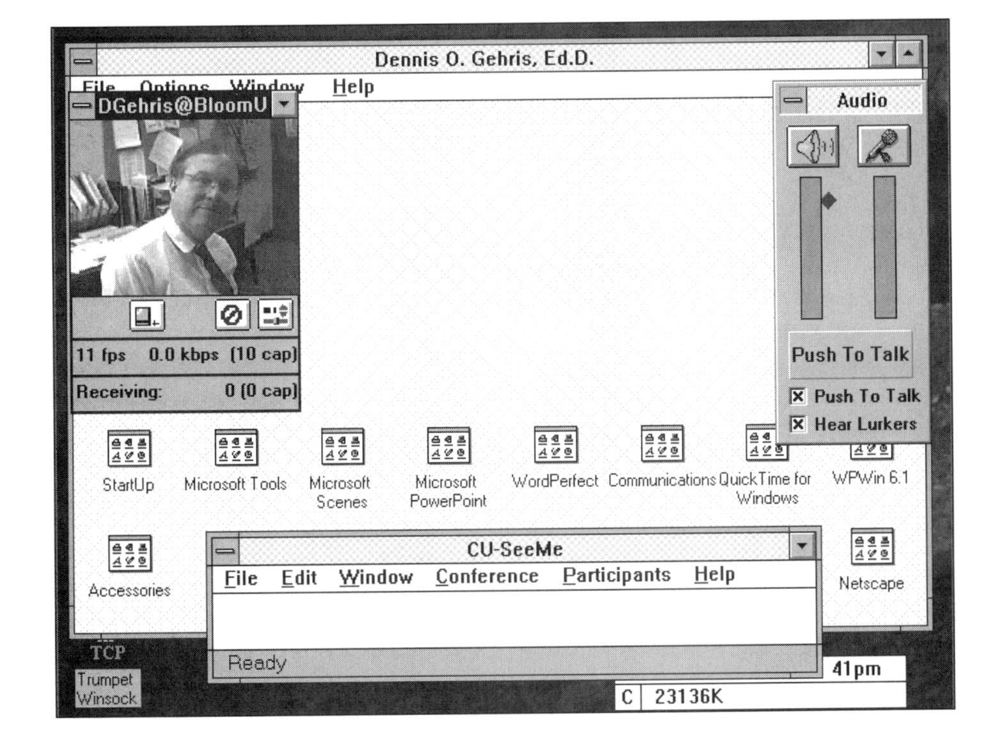

Figure 8-7. The Initial CU-SeeMe Screen

Make the necessary camera adjustments to point the camera at the person or object you intend to send over the Internet. You should also see the Audio dialog box. Check it to see that it is set up properly. The left portion of the box represents settings for the incoming audio. You can adjust the speaker volume by sliding the red marker up and down. The right portion of the box represents settings for your microphone. When you want to speak into your microphone, you need to click and hold in the Push to Talk button. As you speak, the volume bar should move up and down.

Another way to start CU-SeeMe if you're using a PC-compatible computer and Windows is to use CU-SeeMe GO!, which automatically opens CU-SeeMe up to a specific CU-SeeMe site from a Web page. It is also possible to set up a link in your homepage that connects a user to your computer. The URL for the Web page that contains instructions needed to install and configure CU-SeeMe GO! is http://www-personal.umich.edu/~johnlaue/cuseeme/gocusmgo.htm. Linking to http://www-personal.umich.edu/~johnlaue/cuseeme/events.htm will provide an Event Guide that lists current and future CU-SeeMe videoconferencing events. By using CU-SeeMe GO!, you can link to these events by clicking on the IP number or names by using Netscape.

INITIATING A CONNECTION WITH CU-SEEME

To initiate a connection, click on Connect on the Conference menu and the Connect dialog box should appear (see Figure 8-8).

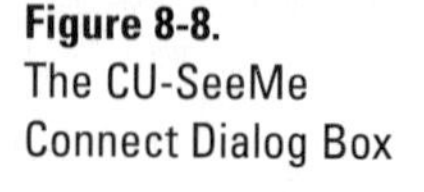

Figure 8-8. The CU-SeeMe Connect Dialog Box

Connect
Enter an IP address or
Connect pro60-test2.cit.cornell.edu
I will send video
I will receive video
Conference ID 0
OK
Cancel

Make sure that both I Will Send Video and I Will Receive Video options are selected. If you know an IP number or name to enter, click an insertion point on the white connect typing area or press the tab key to select the area. Key the IP number or name and click on OK. If you would like to use a previously used IP number or name, click on the down scroll arrow on the typing area. A list of previously entered addresses should appear and you can select one of these. A list of sample IP names may appear. Up to 100 reflector identifiers will appear as you add additional sites to the list.

The following is a sample list of IP names or numbers for active public reflectors in the United States that you might want to try. For a list of currently active reflector addresses, use Netscape to link to http://www-personal.umich.edu/~johnlaue/cuseeme/csmlist.htm. Be aware that many reflectors are not operational 24 hours a day, seven days a week. If you try to connect to a reflector that is not broadcasting, an error message will appear.

Sample List of Active CU-SeeMe Reflectors

Cornell University	pro60-test2.cit.cornell.edu
Digital Jungle	reflector.jungle.com
Educational Computing Network	uxb1.ecn.bgu.edu
Kent State University	business.kent.edu
Miami University (Ohio)	old-holmes.lib.mulhio.edu
NASA at CMU GSIA (receive only)	www.gsia.cmu.edu
NSA TV at Kent State (receive only)	axon.kent.edu
North Carolina State University	magneto.csc.ncsu.edu
NYSERNet	nysernet.org
Ohio State University	davros.acs.ohio-state.edu
Penn State University	hornet.cac.psu.edu
University of Maine	130.111.120.13
University of Maryland	haven.umd.edu
University of Pennsylvania	isis.dccs.upenn.edu
University of Texas	128.83.108.14
Virginia Commonweath University	128.172.157.244

The Conference ID value on the Connect dialog box will normally be defaulted at zero (0). It is a security feature that can be configured by a reflector operator. A reflector will reject all persons attempting to connect if the Conference ID specified doesn't match the ID it has been configured to accept. A zero (0) value will allow all persons connection to the reflector, assuming that there is room.

CONDUCTING A VIDEOCONFERENCE

Once you've made a connection to a reflector, you will see video windows for the number of participants you indicated in the Preferences dialog box, along with your own video window. You are now ready to try to participate in a videoconference. The participant window will show a list of all the participants in the conference. To display the participant window, click on Show Participants on the Participants menu. Senders will be listed in the top section; senders whose windows are currently not showing are listed in the next section; and **lurkers**, users who desire to receive video but are not sending video, are listed in the third (bottom) section. Some reflectors limit the number of lurkers.

Each section of the Participant window can be collapsed by hitting the little button at the left of each section.

Video windows for the senders (up to the maximum you indicate in the Preferences box) will be shown. You can close any video window except your local video window. When a window is closed, that user moves from the Senders section to the second section of the participants list. Clicking on a participant name in the Senders (not showing) section of the participant list will reopen that video window.

At the bottom of each video window are icons that tell you various things about the person. The first one, represented by an eye, shows whether or not the person is viewing you on his or her screen. An open eye indicates that the person sees your video window; a closed eye indicates that he or she does not see your video window. The audio icon enables you to turn off sound from a particular person by clicking until the icon disappears. When the user is transmitting audio, the button is shaded gray. The microphone icon shows a red X if the user cannot transmit sound or has it turned off. The two remaining icons, when clicked, show the participant's transmission details.

On the Windows version of CU-SeeMe, clicking the right mouse button on a video window during a videoconference will bring up a floating pop-up menu containing a list of options that can be set for that video window. The options are as follows: Freeze, which stops the video display for that window; Topz, which keeps that video window on top all of the time so that the window cannot be covered; and Get Info, which displays the name and IP address of that sender.

You can also enter a keyboard message by selecting the local video window and keying a message on your computer keyboard. Doing this causes the characters keyed to be displayed at the bottom of the window and sent with your window. The font face, size, and appearance can be changed.

Figure 8-9 shows a CU-SeeMe videoconference in progress. Note that the Show Participants option has been selected and that the senders and corresponding video windows are displayed; a list of senders (not showing) are also shown. A list of lurkers can be displayed at the bottom of the box. The frames per second (fps) numbers at the bottom of the video windows show how many times per second the video window is updated (redrawn). The kilobits per second (kbps) number shows the speed of the video traveling across the network from that sender. Note that the local video window (DGehris@BloomU) also shows these rates as well as the rate capacity in parentheses and the receiving rate.

You can also chat with conference participants by selecting Open Chat Window from the Conference menu. The chat window provides a way for users to communicate with one another by typing. This is especially useful for those connected to the network with slow (14.4 kbps or less) modems who cannot use audio to communicate. Figure 8-10 shows the CU-SeeMe chat window.

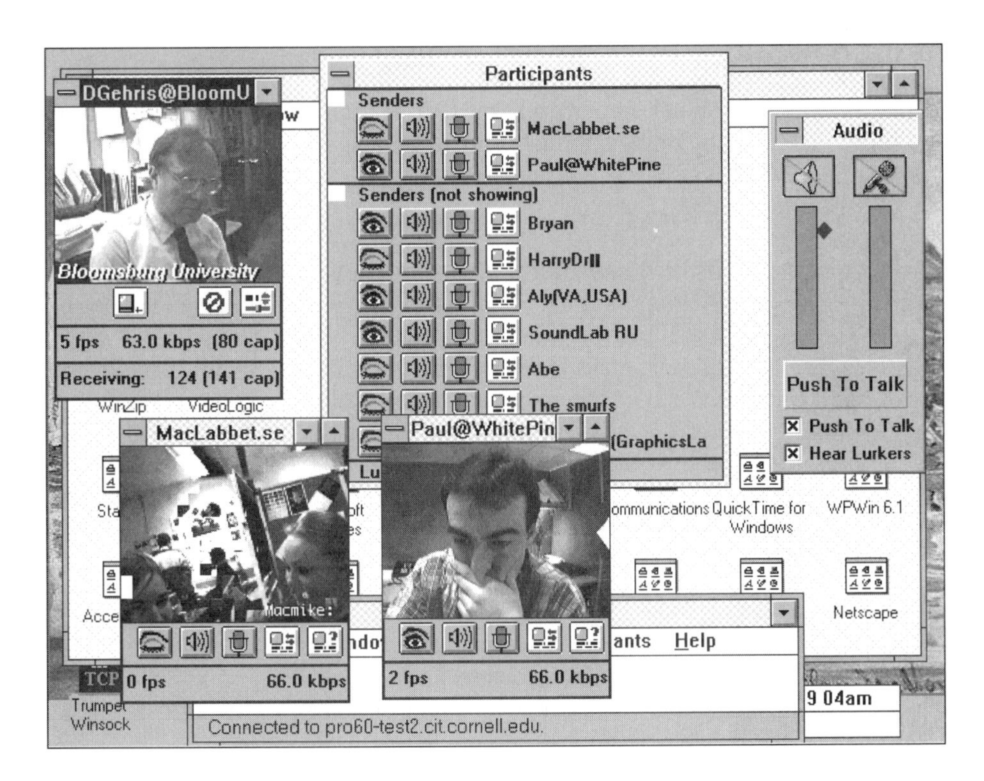

Figure 8-9.
A CU-SeeMe Videoconference in Progress

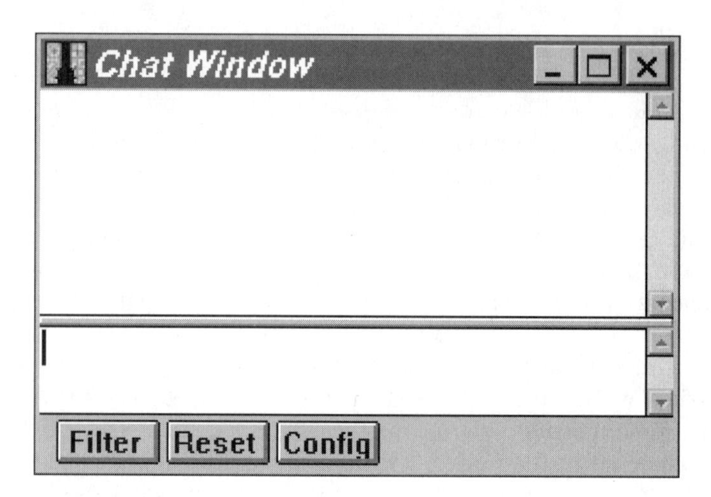

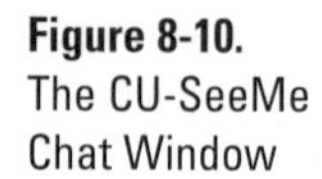

Figure 8-10. The CU-SeeMe Chat Window

Obtaining More Information about CU-SeeMe

There are several ways to get information about CU-SeeMe on the Internet, including mail lists and Web pages.

GENERAL DISCUSSION MAIL LIST

If you'd like information about developments in CU-SeeMe or its use, or in contacting other CU-SeeMe users, an automated mail list has been established. The list is provided for unrestricted discussion of the CU-SeeMe software. To join the list, send an electronic mail message with the following line as the entire message body to listserv@cornell.edu: subscribe cu-seeme-l <your first name> <your last name>, leaving the subject line blank. You should receive a confirming message with extensive instructions on use of the list. You can send mail to be distributed to the list to: cu-seeme-l@cornell.edu. The following is an example of a message in the form of a question that a subscriber sent to this mail list and an answer from another subscriber.

Date: Sat, 03 Feb 96 15:38:59 EST
From: stephen dyason <sdyason@iafrica.com>
To: CU-SEEME-L@cornell.edu
Subject: Re: Can't connect to any hosts.....

On Thu, 01 Feb 1996 17:17:18-0600 you wrote:

>I have CUSEEME and it looks to be a great program but for some reason I
>can never connnect to any hosts. I saw the FAQ but that didn't really
>apply to me. I have Win95, a Static IP address, USR 33.6 modem,

John, tell us what the problem is. Do you get an error message or do you get a busy signal. In other words, is the program malfunctioning, or is it simply bad luck.

sdyason@iafrica.com
OS/2's the answer: WHAT'S THE QUESTION?

ANNOUNCEMENT MAIL LIST

Another mail list is devoted to announcing significant advances in CU-SeeMe technology, new releases of the software, and related issues. To subscribe to Cornell University's CU-SeeMe Announcement List, send electronic mail to listserv@cornell. edu that reads subscribe cu-seeme-announce-l <your first name> <your last name>, leaving the subject line blank.

EVENTS MAIL LIST

A third mail list provides announcements and discussions of broadcasts and events over the Internet that use CU-SeeMe. This list's purpose is to provide information about public broadcasts in which CU-SeeMe users may wish to participate. To subscribe to the CU-SeeMe Events List, send electronic mail to list@www.indstate.edu that says: subscribe cusm-events <your first name><your last name>.

WEB PAGES

Cornell University also maintains a Web page at http://cu-seeme.cornell.edu. An unofficial Web page administered by Michael Sattler is available at http://www.jungle.com/CU-SeeMe. These pages provide information about CU-SeeMe software developments, version announcements, and installation and configuration advice. Figure 8-11 shows the CU-SeeMe Web page at Cornell University.

Enhanced CU-SeeMe

Enhanced CU-SeeMe runs on both Windows and Macintosh computers and offers full-color video, audio, chat window, and whiteboard communications. You can participate in "Live over the Internet" conferences, broadcasts, or chats. Enhanced CU-SeeMe can be launched directly from Web pages with Netscape and other browsers. This version of CU-SeeMe will run with a 28.8k modem, ISDN link, or better. For audio-only telephony use, it works effectively over a 14.4k modem.

Enhanced CU-SeeMe uses a unique protocol to manage, receive, and rebroadcast video and audio data. The protocol was developed specifically for TCP/IP networks and the Internet. Person-to-person, group conferencing, and large audience broadcasting over TCP/IP networks are all possible.

GENERAL FEATURES

Enhanced CU-SeeMe will run on the following platforms: Windows 3.1, Windows 95, Macintosh, and Power Macintosh. You are able to view up to eight participant windows, with an unlimited number of participants for the audio and talk window. The program features a Caller ID that is a message alert box

Figure 8-11. The CU-SeeMe Web Page at Cornell University

for incoming connections. A whiteboard exists for collaboration during conferences that supports multiple users. Color is supported with 24-bit true color or 4-bit greyscale. A phone book permits you to save, add, and edit participant addresses and reflector sites. A demonstration version of Enhanced CU-SeeMe is available at http://www.wpine.com/cudownload.htm. This is a fully functional version of the product and does not require a serial number. However, the program will time out after 30 minutes and will need to be restarted. In addition, it will remain active for a limited amount of time. See Figure 8-12 for an example of an Enhanced CU-SeeMe videoconference in progress.

Comparison of Web-Based Conferencing Application Programs

Now that we have reviewed computer conferencing and videoconferencing applications in Chapters 7 and 8, it is time to compare the application programs. Table 8-1 provides a cross-matrix of the various media formats that each of the Web-based conferencing application programs offer in terms of support for chat (text), graphics, audio, video (including animation), whiteboard, and answering machine formats. It should be noted that Netscape Chat provides graphics through the display of the Netscape browser.

Figure 8-12. An Enhanced CU-SeeMe Videoconference in Progress

TABLE 8-1 CROSS-MATRIX OF MEDIA FORMATS PROVIDED BY WEB-BASED CONFERENCING APPLICATION PROGRAMS

Applications	Chat (Text)	Graphics	Audio	Video	White-board	Answering Machine
Netscape Chat	X	X				
WebChat	X	X	X		X	
Internet Phone			X			
Cool Talk	X		X		X	X
CU-SeeMe	X		X	X	X	
Enhanced CU-SeeMe	X		X	X	X	

The second comparison we will make is in communication modes among the various conferencing application programs, as shown in Table 8-2. The three types of communication modes are: (1) one person communicating to one other person, (2) one person communicating to many individuals, and (3) many persons communicating to many others with each transmission. As you can see, CU-SeeMe is the only application we have reviewed that provides the opportunity to communicate via all three modes.

TABLE 8-2 CROSS-MATRIX OF COMMUNICATION CHARACTERISTICS PROVIDED BY WEB-BASED CONFERENCING APPLICATION PROGRAMS

Applications	One-to-One	One-to-Many	Many-to-Many
Netscape Chat	X	X	
WebChat	X	X	
Internet Phone	X		
CoolTalk	X		
CU-SeeMe	X	X	X
Enhanced CU-SeeMe	X	X	X

Questions for Review

1. What is videoconferencing? How does it differ from computer conferencing, business television, and audio conferencing?
2. What barriers to the use of videoconferencing exist today? How will these problems be solved in the future?
3. What different types of facility configurations are available for videoconferencing?
4. What are possible videoconferencing applications for business, education, and personal use?
5. What tips for conducting a successful videoconference can you list for lighting, cue cards/eye contact, audio, photos and props, interactivity, appearance and attention, moderators, video clips, and rehearsals?
6. What is the purpose of CU-SeeMe software? What are the hardware requirements for using CU-SeeMe?
7. How is CU-SeeMe software downloaded, configured, and started?
8. How do you initiate a connection and conduct a CU-SeeMe videoconference?

9. How can you obtain additional information about CU-SeeMe?
10. How does the enhanced version of CU-SeeMe differ from the regular version?
11. How do the computer conferencing and videoconferencing application programs reviewed in Chapters 7 and 8 compare in types of media formats?
12. How do the computer conferencing and videoconferencing application programs reviewed in Chapters 7 and 8 compare in communication characteristics?

Exercises

1. If necessary, download and install CU-SeeMe software. Use the URLs provided in the chapter.
2. Use CU-SeeMe to initiate a point-to-point videoconference with another person who also has installed CU-SeeMe software and has a connected video camera for 15 minutes or more. Make sure that you know the IP address of the other computer. Write a paragraph describing the topic(s) discussed and the benefits you see for videoconferencing in this manner.
3. Use CU-SeeMe to participate in a group conference by connecting to a public reflector for 15 minutes or more. See the sample list of reflectors in the chapter. Try to communicate with one or more of the other participants. Write a paragraph describing the topic(s) of discussion and the benefits you see for videoconferencing in this manner.
4. Use CU-SeeMe to participate in a one-way broadcast to one of the NASA TV reflectors and connect to the site for 15 minutes or more. Make sure you deselect the I Will Send Video option on the Connect dialog box. Write a paragraph describing what was broadcast and the benefits that you see for videoconferencing in this manner.
5. Subscribe to the CU-SeeMe general discussion mail list. Send an electronic mail message in which you ask a question about CU-SeeMe. Check your mail to determine if your question was answered. Submit a printout of your question and answer(s).
6. Subscribe to the CU-SeeMe announcement and events mail lists. Did you receive any electronic mail messages from these lists? If you did, submit printouts of the electronic mail messages.
7. Use Netscape to link to one of the CU-SeeMe Web pages. What information is contained on these pages?

8. Study Figure 8-9 and answer the following questions:
   a. What is the name of the reflector to which participants are connected?
   b. What are the names of the two senders for which video windows can be seen?
   c. Which one of these senders can see our video window? Which one cannot see our video window?
   d. How many frames per second will *Paul@WhitePine*'s video window be updated? How many times per second will *DGehris@BloomU*'s video window be updated?
   e. How many kilobits per second is both *MacLabbet.se*'s and *Paul@WhitePine*'s video window traveling across the Internet? What is the kilobits per second rate for *DGehris@BloomU*'s video window?
   f. Will *DGehris@BloomU*, the local video window, be able to hear audio transmissions? Will he be able to transmit audio? Why or why not?
9. Study Figure 8-12 and answer the following questions:
   a. What is the difference between this figure and Figure 8-9?
   b. What are the names of the three senders for which video windows can be seen?
   c. Which one of these senders can see our video window? Which one cannot see our video window?
   d. What are the names of the hidden users?
   e. What are the names of the lurkers?
   f. What would Useless like the other participant visit?
   g. Why don't we see Eva of *Swed.de.me*'s video window, even though she is chatting? Can the other participants talk to her via audio? Why or why not?

Case

The Louiston Insurance Company, which sells health insurance plans to large- and medium-sized companies in a seven-state region in the Midwest, currently conducts two-day training sessions for its fifteen regional sales managers each month at the company's headquarters in Chicago. Louiston reimburses each of these individuals for round-trip airfare, which averages $240 each month. Additional expenses amount to $170 per person per month for hotel accommodations, and $150 per person for meals. These two-day training sessions are used to provide managers with information on everchanging insurance regulations, insurance policy provision changes, and new sales techniques. Louiston is considering videoconferencing for alternating months (six times per year) in lieu of the face-to-face sessions. The company would retain traditional sessions for the other months. Write a position statement and at least two supporting paragraphs describing why you believe the company should or should not make this change. Is cost saving the only consideration? What would be the benefits of using videoconferencing? What would be the drawbacks?

CHAPTER

9

Adobe Acrobat, ASAP WebShow, and Macromedia Shockwave on the Internet

This chapter introduces three multimedia programs: Adobe Acrobat, ASAP WebShow, and Macromedia Shockwave. Adobe Acrobat and Macromedia Shockwave are available for both the IBM and IBM PC–compatible and Macintosh platforms, and ASAP WebShow is available only for IBM and IBM PC–compatible computers at the writing of this textbook. You will learn the functions of all three applications as well as the steps and procedures for using Adobe Acrobat to read those documents on the Web that use the PDF file format.

What You Will Learn

- How Adobe Acrobat is defined
- Advantages of using Adobe Acrobat PDF files
- Steps and procedures for viewing Adobe Acrobat PDF files
- Parts of the Adobe Acrobat reader

- Examples of Web sites that use the PDF file format
- How ASAP WebShow is defined
- Characteristics of ASAP WebShow
- Examples of Web sites that use the ASAP file format
- How Macromedia Shockwave is defined
- Advantages of using Macromedia Shockwave
- Steps and procedures for downloading and installing the Shockwave Plug-in
- How to access sites that use Shockwave

Adobe Acrobat Defined

Adobe Acrobat is a document format that includes capabilities for retaining original page layout information, fonts, graphics, and the like. Although you need to purchase the commercial Acrobat software to create Acrobat pages, there is no charge to configure Acrobat as a helper or plug-in application in Netscape and other browsers. The Netscape plug-in technology enables third-party developers and users to extend the capabilities of the Netscape Navigator. To use the helper or plug-in application, you need to find Web sites that provide files in the **portable document format** (**PDF**). Figure 9-1 shows a PDF file that has been downloaded using Netscape and the Acrobat Reader.

Advantages of Using Adobe Acrobat PDF Files

One advantage of using portable document format files is that they are universal, so that almost anyone (Windows, Macintosh, DOS, and UNIX system users) can view and print the files. There is no difference in the appearance of the files when they are viewed and printed with Netscape, Mosaic, or any of the other browsers. Another advantage is that the originator is not limited to standard HTML formatting and can retain the original document formatting, including scaleable typefaces and graphics. Finally, it is possible to integrate HTML files into PDF files so that links can be included to guide readers through homepages, documentation, and to other Web pages.

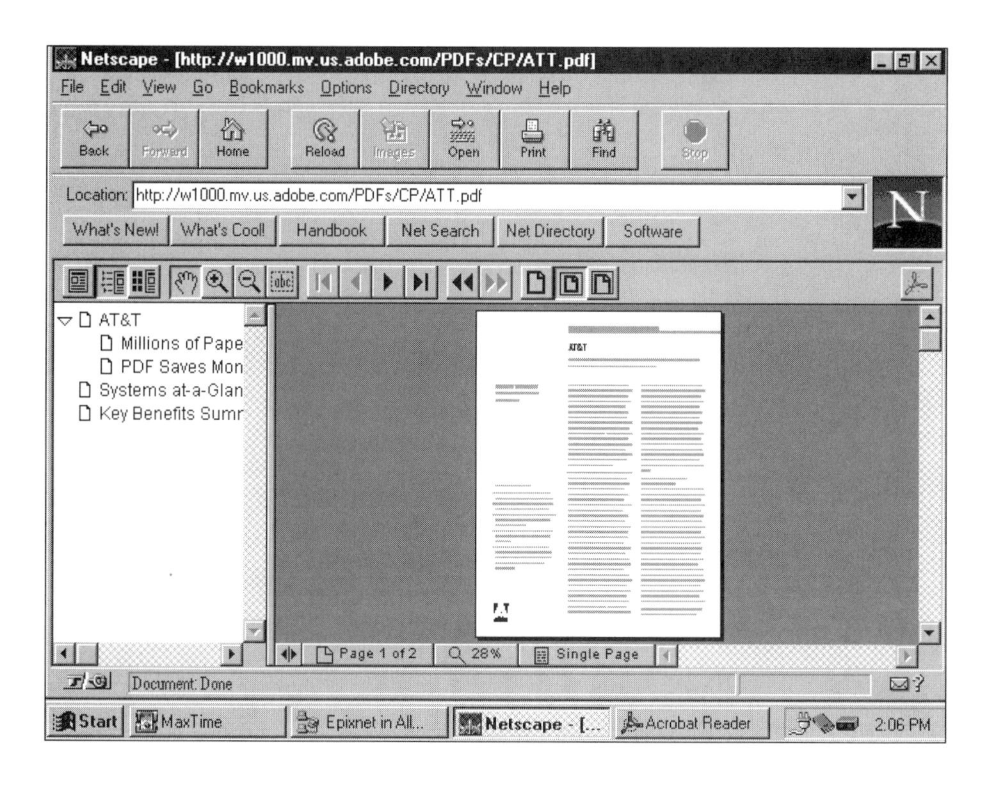

Figure 9-1. A PDF Sample Page

Steps and Procedures for Viewing Adobe Acrobat PDF Files

Here's how to view Adobe Acrobat PDF files.

DOWNLOADING THE VIEWER

To view Adobe Acrobat (PDF) files with Netscape, you must first download the viewer. Using Netscape, open the Adobe Acrobat homepage at http://www.adobe.com/Acrobat. (Figure 9-2 shows this page.) You need to click on one of the Adobe Acrobat Reader links and follow the onscreen instructions. You will save the reader in a subdirectory of your hard disk so that it can be accessed when you are setting up the reader as a helper application. The reader can be used as a standalone program if you happen to have a PDF file that you download or have created it using the Adobe Acrobat creation program.

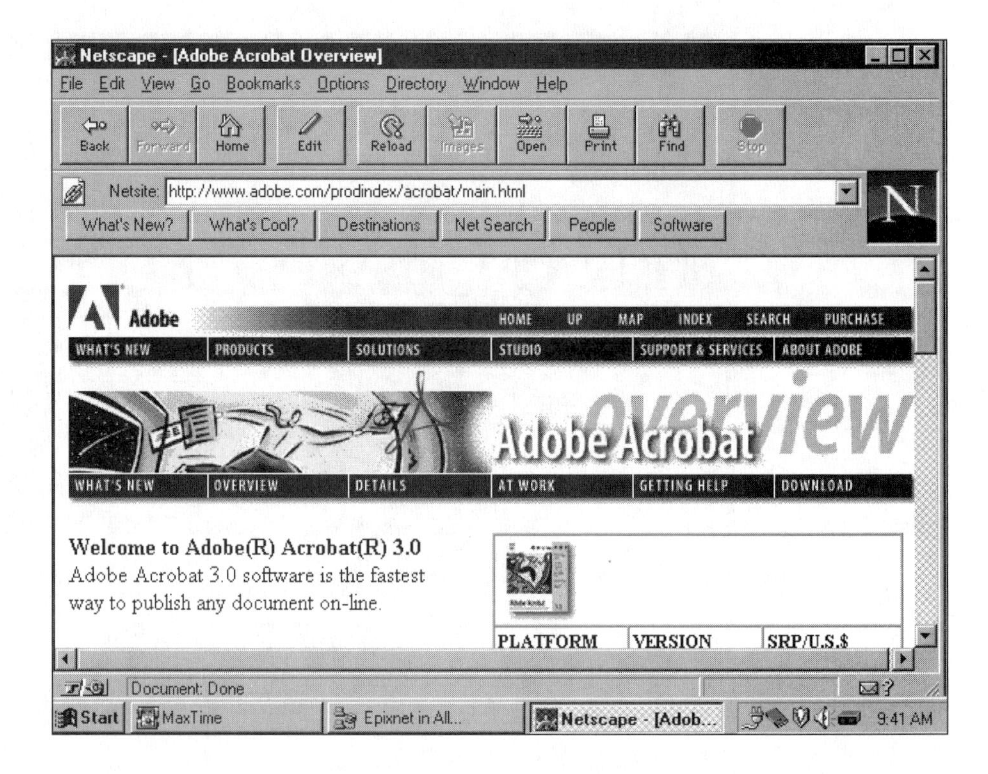

Figure 9-2. Adobe Acrobat Homepage

CONFIGURING NETSCAPE FOR ADOBE ACROBAT

Configuring the Adobe Acrobat reader as a Netscape 3.0 helper application for either the PC-compatible or Macintosh computers involves the following steps.

1. Select General Preferences on the Options menu. Click on the Helpers tab to display the Preferences dialog box. A list of MIME types appears, along with controls to specify the action the browser should take when it encounters a file of a specific type.
2. Click on the Create New Type button. The Configure New MIME Type dialog box appears, with typing boxes for MIME Type and MIME Sub Type.
3. Key application in the MIME Type box.
4. Key pdf in the MIME Sub Type box. Click on OK. The Configure New MIME Type dialog box will disappear.
5. Key pdk in the Extensions typing box.
6. Click on the Launch Application radio button and enter the path and filename of your Acrobat reader. If you do not know the path of your reader, press the Browse button and use the controls of the dialog box that appears to select the path. Compare your screen with Figure 9-3. Click on OK.

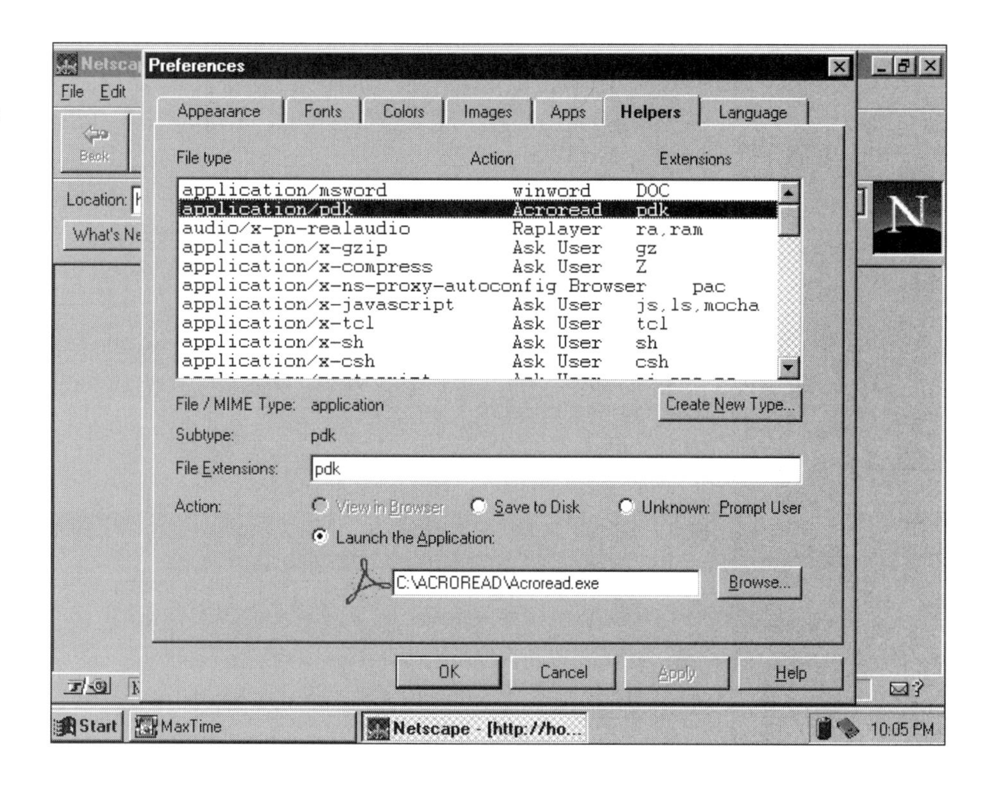

Figure 9-3. Adobe Acrobat Reader Added as Netscape Helper Application

Netscape will now use the Acrobat reader as a helper application when you click on the link of a PDF file.

Linking to a PDF File on the Web

After you have downloaded the Adobe Acrobat reader and have configured it as a helper application in Netscape or some other browser, you are ready to link to a PDF file on the Web. As an example of linking to a PDF file, follow these steps using an IBM or IBM PC–compatible computer.

1. Use Netscape to open http://www.adobe.com/studio/spotlights/main.html. #acrobat. This is a page containing a number of sample PDF files.
2. Click on the icon at the left of the item labeled Customer Trends: Worldwide Publishing with Adobe Acrobat Software (704 KB/6 pages). This file contains information relating to integrating Acrobat into the Web and discusses how the government, corporations, educational institutions, and professional publishers are using Acrobat. You may want to print this file after you've downloaded it. If this document is not available, choose another one.

3. The PDF file will be downloaded and saved in a temporary directory on your hard disk (probably c:/netscape/cache or c:/program files/netscape /navigator/cache). The Acrobat reader will open and the introductory screen will appear.
4. The file should open from within Netscape. If the file does not appear, click on Open File on the File Menu. Find the PDF file (M0ok990g.pdk or something similar) in the cache subdirectory, using one of the paths in step 3.
5. Click on the Bookmark/Page Display button on the toolbar to view the bookmarks and the page display. This is the second button from the left. (Your screen should look like Figure 9-4.)

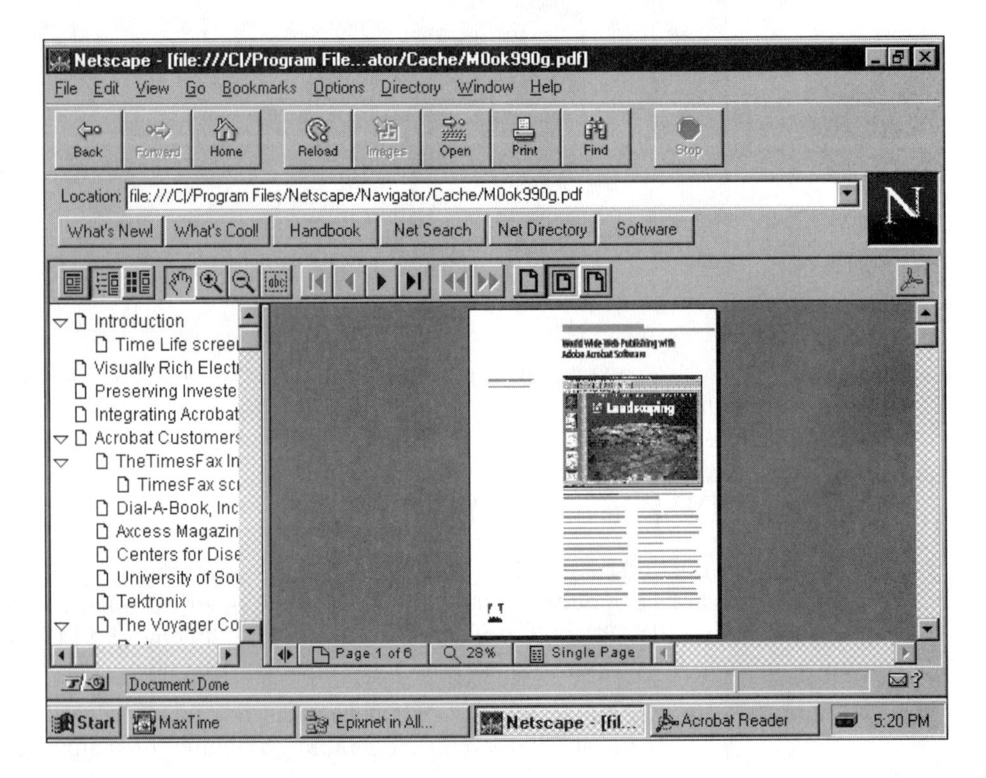

Figure 9-4. Adobe Acrobat Reader with Open File

Identifying the Parts of the Acrobat Reader

The Adobe Acrobat reader consists of the following parts, as shown in Figure 9-5.

MENUS

The menus display application commands, many of which can also be executed by using one of the buttons on the toolbar. The File menu is used to open, close, and print PDF documents as well as providing document information. The Exit item is used to exit the program. The Edit menu contains an item to undo a previous command. The Cut, Copy, Paste PDF Document places items into and from the Windows clipboard and the Preferences items are used to make configuration changes. The View menu items allow the user to display files in various ways. The Tools menu contains items for zooming in and out for different views and also allows the user to select text and graphics and to find selected text. The Window and Help menus are standard Windows menus for adjusting how windows are displayed and providing a means for obtaining help on using the reader. The Help files need to be downloaded separately from the reader and may or may not be available.

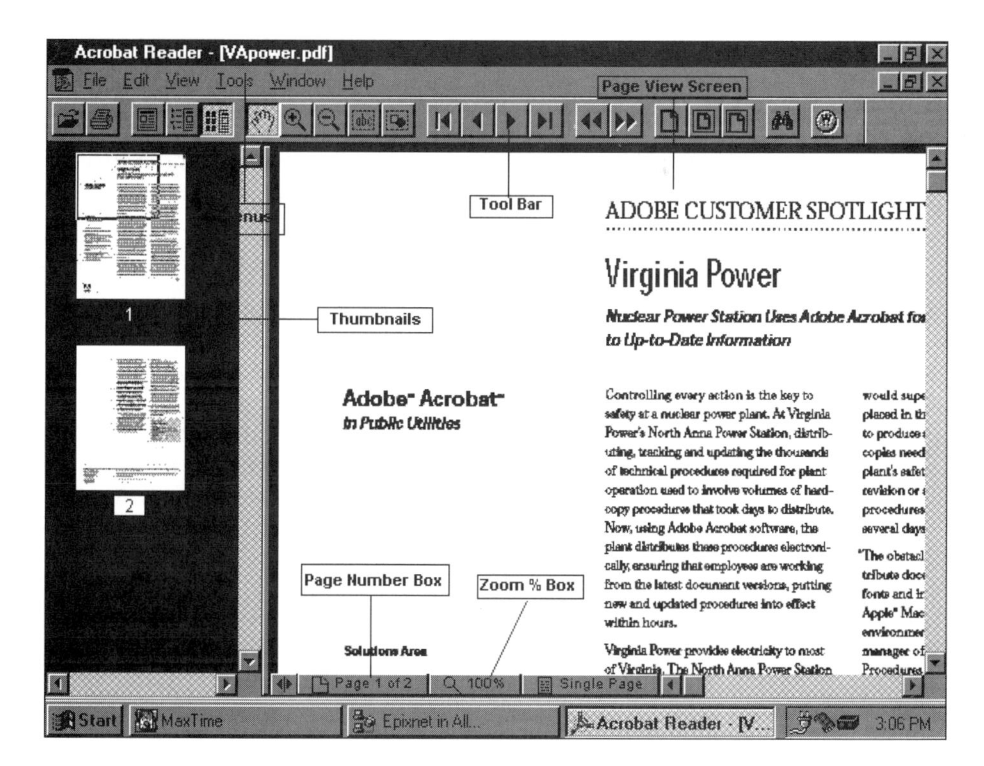

Figure 9-5. Parts of the Acrobat Reader

TOOLBAR

The toolbar is used as an alternative method of accessing application commands. You can remove the toolbar from the application window to provide more room in the client area to display documents. To hide the toolbar, select Hide Toolbar from the Window menu. To display the toolbar again, select Show Toolbar from the Window menu. Use Figure 9-6 to identify each button on the toolbar by letter. Some of the buttons will be slightly different on later versions of Adobe Acrobat.

A *Open* — Used to open a PDF file for display. After clicking on the button, you will be permitted to indicate the path and filename for the file you want to open. An alternate way of opening a PDF file is to choose Open on the File menu.

B *Print* — Used to print a PDF file, which is displayed in the application window. A Print dialog box will appear in which you can indicate the number of copies and pages to be printed and the desired print quality. An alternate way of printing a PDF file is to choose Print on the File menu.

C *Page Display* — Displays only the current page of the PDF document.

D *Bookmark/Page Display* — Displays the bookmark at the left side of the window and the page display at the right. See later for a description of bookmarks.

E *Thumbnail/Page Display* — Displays thumbnails at the left side of the window and page display at the right. See "The Thumbnails" later in this chapter for a more complete description.

F *Hand* — Transforms the mouse pointer into the hand tool when it is over a document window.

G *Zoom In* — Transforms the mouse pointer into the Zoom In tool when it is over the active document window's page view screen. Click the tool over an active page view to double the view's magnification level up to 800 percent, the maximum magnification level.

H *Zoom Out* — Transforms the mouse pointer into the Zoom Out tool when it is over the active document window's page view screen. Click the tool over an active page to halve the view's magnification level down to 12 percent, the minimum magnification level. An alternate way of zooming out is to choose Zoom Out on the Tools menu.

I *Text Selection* — Transforms the mouse pointer to the text selection tool (I-beam) when it is over the active document window's Page View screen. Holding down the mouse button and dragging the Text Select tool over an active Page View selects text. The text can be copied to the clipboard using the Edit menu's Copy command.

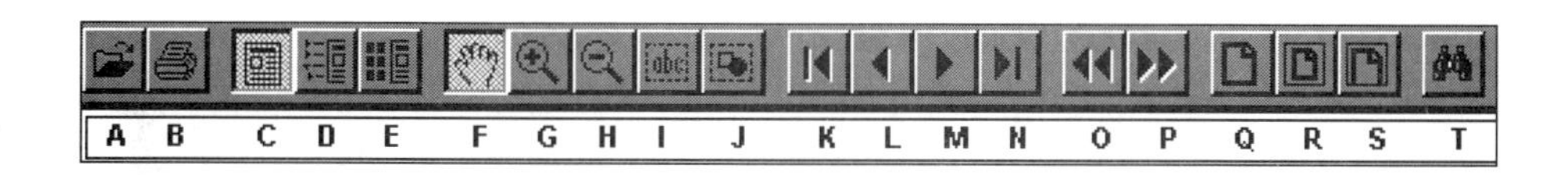

Figure 9-6. The Adobe Acrobat Reader Toolbar

J	*Graphics Selection*	Changes the mouse pointer to the Graphics Selection tool (crossbar) when it is over the active document window's Page View screen. Holding down the mouse button and dragging the Graphics Selection tool over an active Page View draws a Graphic Selection box. An alternate way of selecting the Graphics Selection tool is to choose Select Graphics on the Tools menu.
K	*First Page Display*	Displays the first page of the PDF document if another page is currently in the Page View screen.
L	*Previous Page Display*	Displays the previous page of the PDF document if a page other than the first page is in the Page View screen.
M	*Next Page Display*	Displays the next page of the PDF document in the Page View screen.
N	*Last Page Display*	Displays the last page of the PDF document in the Page View screen.
O	*Previous View*	Changes the Page View screen to the previous view selected.
P	*Return to Next View*	Returns the Page View screen to the view before the previous View button was selected.
Q	*100% Zoom*	Displays the PDF document in the Page View screen to 100% view.
R	*Current Page Fit Inside Window*	Displays the PDF document in the Page View screen to full page view.
S	*Visible Width of Current Page Fit Inside Window*	Displays the PDF document in the Page View screen to show the complete width of the page.
T	*Find Specified Text*	Displays the Find Dialog box. You can then specify a word or phrase and click on Find to determine if it exists in the displayed document. If the word or phrase is found, the first occurrence will be highlighted. You may need to select a larger view to read the selection. Clicking on the Find Specified Text button again will permit you to search for additional occurrences. Clicking on Cancel will allow you to specify a new word or phrase. There may be additional buttons on the toolbar relating to searching text. (Note: This button may not appear on the toolbar displayed in Netscape after downloading a PDF file.)

THE PAGE VIEW SCREEN

The Page View screen (sometimes referred to as the Page View pane) is the area within the Acrobat reader window that displays page views of the PDF document that has been opened.

THE THUMBNAILS

Located on the left side of the Acrobat reader window when displayed, **thumbnails** are miniature graphical representations of a document page. The thumbnail for the active document visible in the Page View screen also has a view box that you can resize and drag to other positions on the same page by using the hand tool. To move the view box to another thumbnail, click the desired page's thumbnail and the view box moves to the thumbnail, appearing centered over the click point if possible. To resize a view box, drag the view box's resizing handle to a different location. When a thumbnail is completely gray, the thumbnail does not exist. (See Figure 9-5 for an example of thumbnails.)

THE BOOKMARKS

Bookmarks, when displayed at the left side of the Acrobat reader window, allow you to display Page View destinations by clicking on a page icon to the left of the Bookmark name. Bookmarks may appear in any bookmark list position independent of their associated document destinations. Using optional indentation, a document's bookmark list can collectively be a hierarchical outline of the document's content. A bookmark may be subordinate to other bookmarks and have subordinate entries to itself. (See Figure 9-4 for an example of bookmarks.)

THE PAGE NUMBER BOX

This box indicates the Active Page View's page number. Clicking the page number box produces a dialog box that enables you to enter a new page number.

THE ZOOM % BOX

This box indicates the current Page View's zoom percentage. Clicking on the Zoom % box produces a list of possible views. Clicking on any of them will produce the new view. To the right of this box is the number of the Page/Column view box that permits you to change the number of pages (or columns) appearing in the Page View screen.

Examples of Web Sites Using the PDF File Format

An increasing number of organizations use the PDF file format to display pages on the Web. The following is a partial list of organizations and corresponding URLs for PDF documents on the Web at the time this text was written. Open http://www.adobe.com/studio/spotlights/main.html#acrobat for a current list of sites. This list, maintained by Adobe, is updated frequently. New entries are added to the bottom of the list. You can also use a search engine to search on PDK or Acrobat to find additional sites that use PDF files.

Acropolis, the magazine of Acrobat publishing
http://www.acropolis.com/acropolis/

AT&T Technology Magazine
http://www.att.com/att-technology/index.html

Bureau of Economic Analysis
http://www.doc.gov/resources/bea/uguide.html
http://www.doc.gov/resources/bea/otherpub.html
http://www.doc.gov/resources/bea/scbinf.html

Hot Lava Magazine
http://www.interverse.com/lava/13/contents.html

The Smithsonian Institution
http://www.si.edu/reader/acrobat.htm

TimesFax Internet Edition from *The New York Times*
http://nytimesfax.com

U.S. Internal Revenue Service's tax forms page
http://www.irs.ustreas.gov/prod/forms_pubs/forms.html

U.S. Postal Service
http://www.usps.gov/business/

ASAP WebShow Defined

The **ASAP WebShow** application allows you to view, download, and print graphically rich reports and presentations from the Web that were created by Software Publishing Company's WordPower software. WebShow is a Netscape Navigator plug-in presentation viewer for Version 2.0 and above and will install automatically as a part of the installation procedure. It can also be installed as a helper application for Microsoft Internet Explorer and other browsers by associating the .ASP file extension to application/x-asap. At the writing of this textbook, the WebShow viewer was only available for IBM and IBM PC–compatible computers using Windows 3.1 or Windows 95. It can be downloaded by linking to http://www.spco.com/PRODUCTS/WSMAIN.htm.

Characteristics of ASAP WebShow

What are some of the characteristics of WebShow? First, it lets you listen to sound and voice while viewing WebShow presentations on the Web by using RealAudio (see Chapter 5) in conjunction with WebShow. Second, you can view presentations in a small live area window (see Figure 9-7), enlarged to fill the current Web page (see Figure 9-8), or zoomed to full screen (see Figure 9-9). Third, over a dozen border styles and transition effects are available to refine the presentation delivery. Finally, WebShow ASAP files are very small in size and can therefore be downloaded from the Internet in very short periods of time.

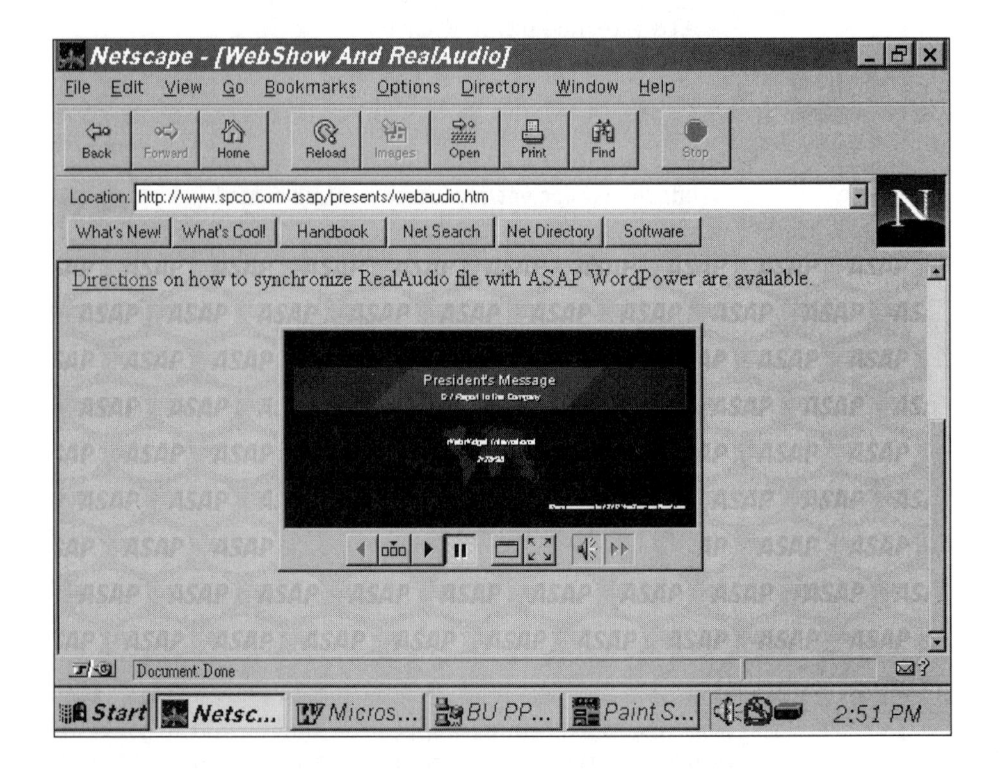

Figure 9-7. WebShow Presentation in Small Live Area Window

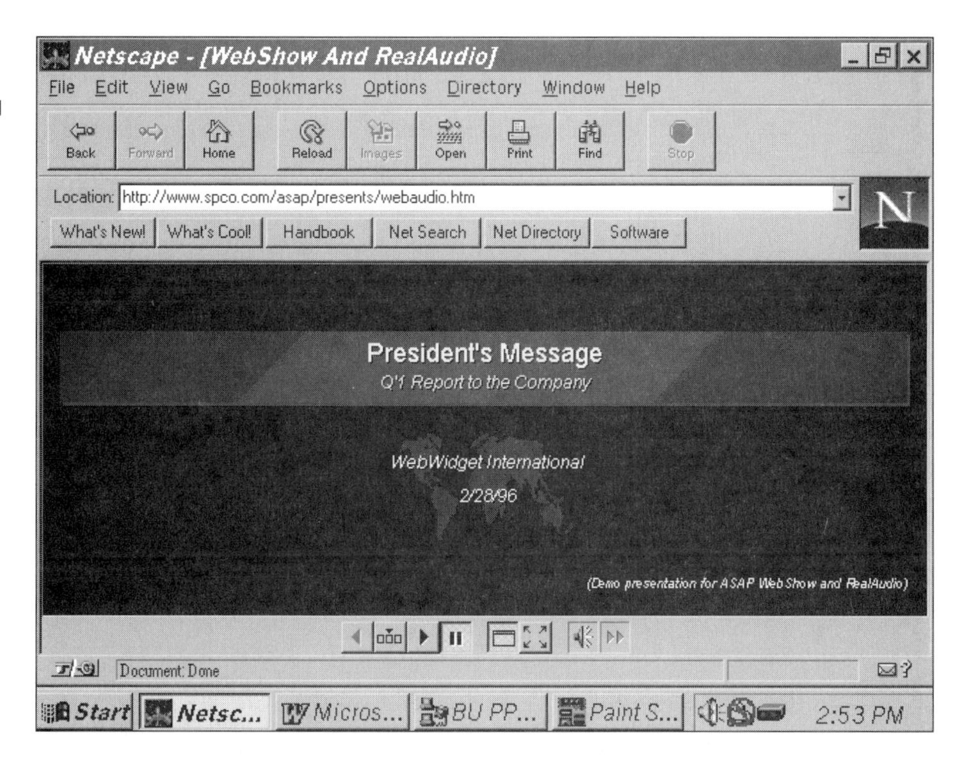

Figure 9-8. WebShow Presentation Filling Current Web Page

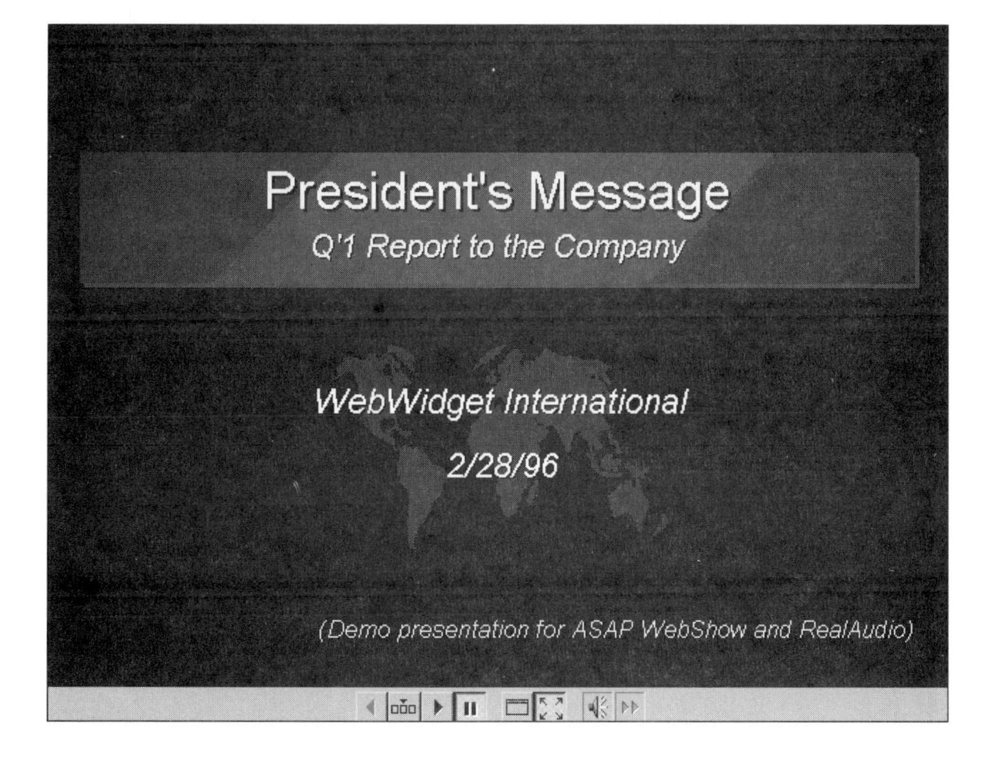

Figure 9-9. WebShow Presentation Zoomed to Full Screen

Examples of Web Sites that Use the ASAP File Format

The following are examples of Web sites that utilize the WebShow ASAP file format.

PC WORLD

PC World magazine uses WebShow to illustrate why they believe they are the only place you need to go for answers to all of your personal computing needs (see Figure 9-10). The URL for PC World's WebShow page is http://www.pcworld.com/resources/promos/asap-presentation.html.

GRAPHICSLAND

Graphicsland, a leading producer of 35mm slides created from ASAP WordPower, is using ASAP WebShow to promote its new online business (see Figure 9-11). The URL for the Graphicsland WebShow page is http://www.graphicsland.com/asapshow.htm.

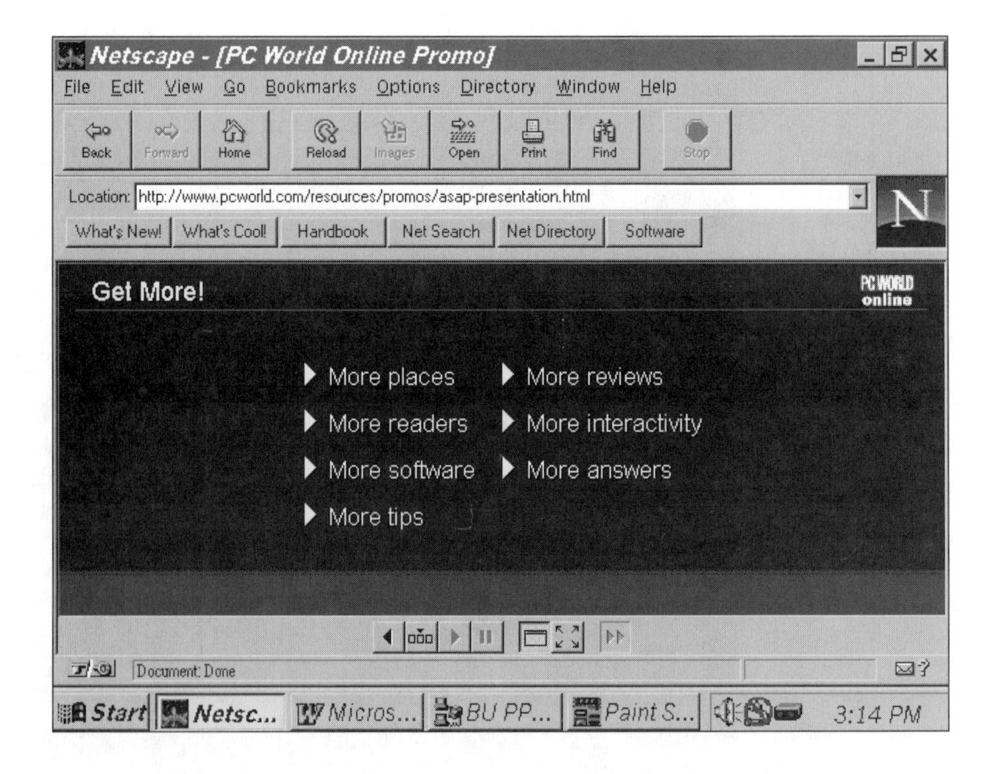

Figure 9-10. PC World's WebShow Presentation

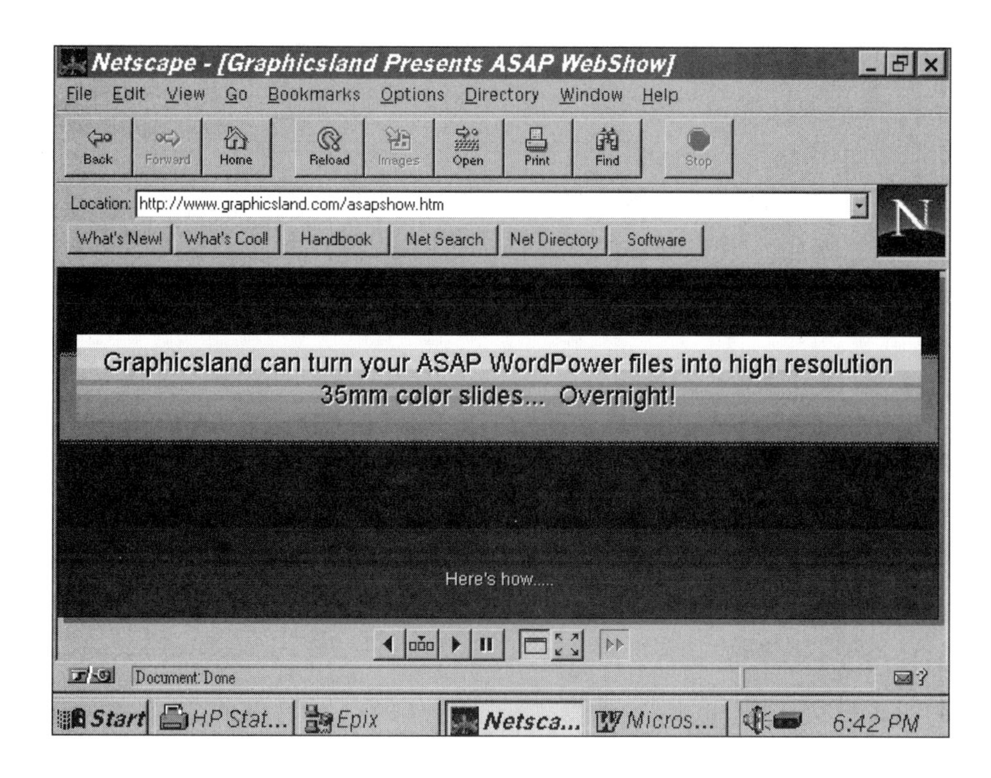

Figure 9-11. Graphicsland's WebShow Presentation

Macromedia Shockwave Defined

Shockwave is a way of viewing interactive multimedia and high-impact graphics on the Web. It is installed in Netscape and other browsers as a plug-in so that files produced with Macromedia Director software are viewable. See page 213 in Chapter 10 for an explanation of plug-in modules (plug-ins). Macromedia Director is a multimedia authoring program that includes an easy-to-learn yet powerful scripting language called Lingo. More than 250,000 multimedia developers use Director software, and major companies such as General Motors, Intel, Sony, MCI, CNN, and Paramount use it to produce their Web pages. Figure 9-12 shows the Intel Web site that uses Shockwave.

Figure 9-12.
Intel Web Site
Using Shockwave

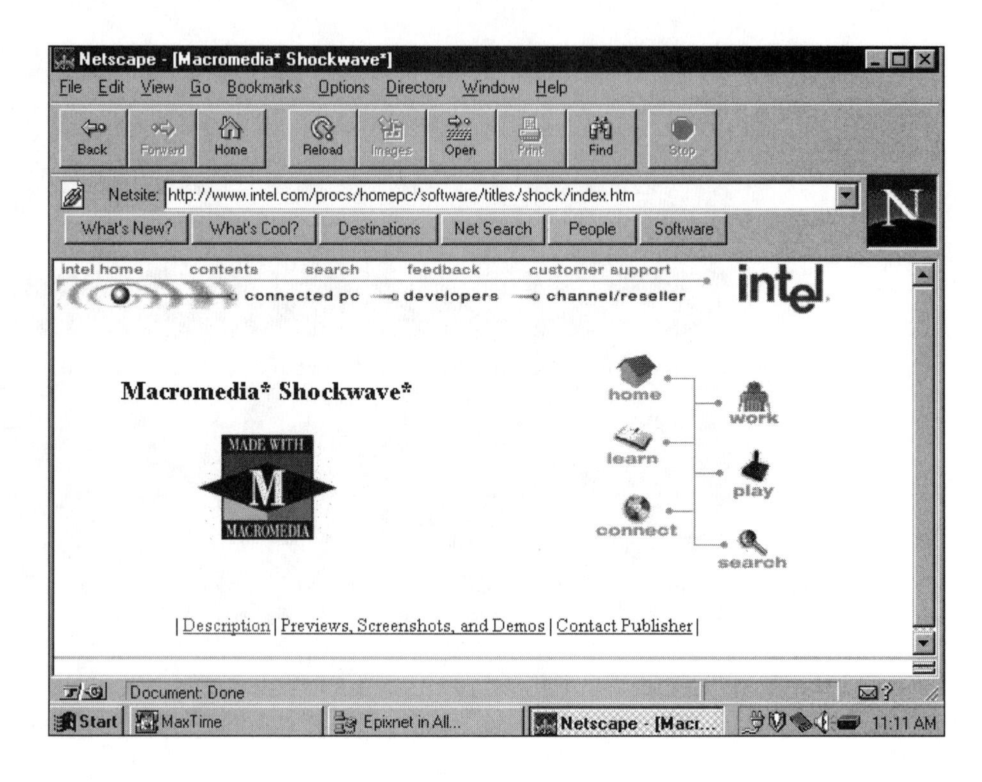

Advantages of Using Macromedia Shockwave

The advantages are numerous. First, Shockwave is becoming a standard for users who want to incorporate motion and interactivity into their Web pages. Second, because Shockwave is so widely used, you are able to reach a massive audience. Third, it is HTML compliant—that is, files with DIR or DCR file extensions are recognized as a part of HTML, using the <embed> tag. Fourth, Shockwave uses the Director engine, which has been proven the best for producing quality multimedia. Fifth, Shockwave adds rich media to Web pages without using a large amount of bandwidth. Sixth and last, Shockwave makes it possible to incorporate new applications, such as real-time games, puzzles, and graphing, into Web pages.

Steps and Procedures for Downloading and Installing the Shockwave Plug-in

To view a Web site that uses Shockwave, you must download the Shockwave plug-in that works with Netscape. Follow these steps:

1. Use Netscape to link to http://www.macromedia.com/shockwave/download.
2. Complete the form that appears on the page (see Figure 9-13). Select the package you wish to download and clock on the Get Shockwave... button.
3. A list of sites where you can download Shockwave will appear. Click on the name of one of the servers listed. If the Server you choose is busy, choose a different one from the list. If all the servers are busy, wait a few minutes and try again. As the Shockwave package begins to download, a Save dialog box should appear.
4. Choose the directory where you want to save the Shockwave Plug-in package. Your browser is located in the default directory; you may choose a different one if you wish.
5. Click on Save.
6. Exit Netscape.
7. Locate the Shockwave file that you downloaded.

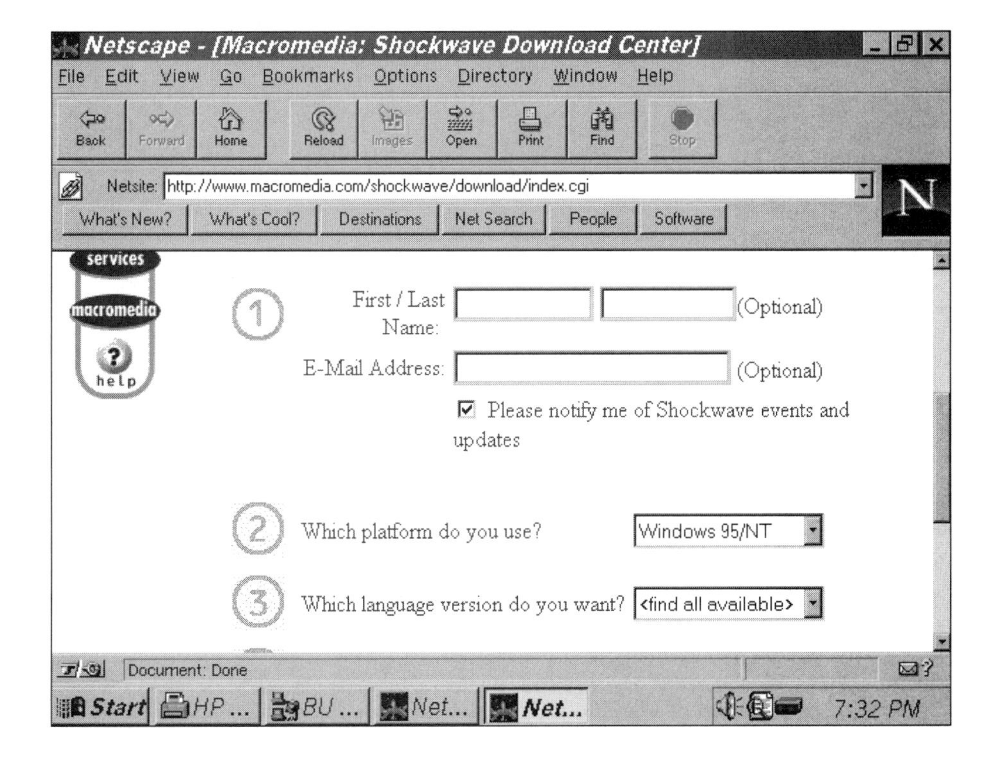

Figure 9-13. Shockwave Download Form

8. Choose Run to run the file that will launch the installer. Follow the instructions that appear on the screen.
9. Restart Netscape.
10. Click on About Plug-ins on the Help menu. Shockwave should be listed as one of your installed plug-ins.

Accessing Web Sites that Use Shockwave

An excellent listing of Web sites that use Shockwave is found on the Shockwave Gallery. You can link to this page by using http://www.macromedia.com/shockwave/epicenter/index.html (see Figure 9-14). This site shows the best "shocked" sites. For example, selected sites are listed as the "Shocked Site of the Day & Week." Sites are also categorized under the following headings; Corporate, Design, Education, Entertainment, Illustration & Cartography, Internet, Intranet, Japanese, and Mass Media.

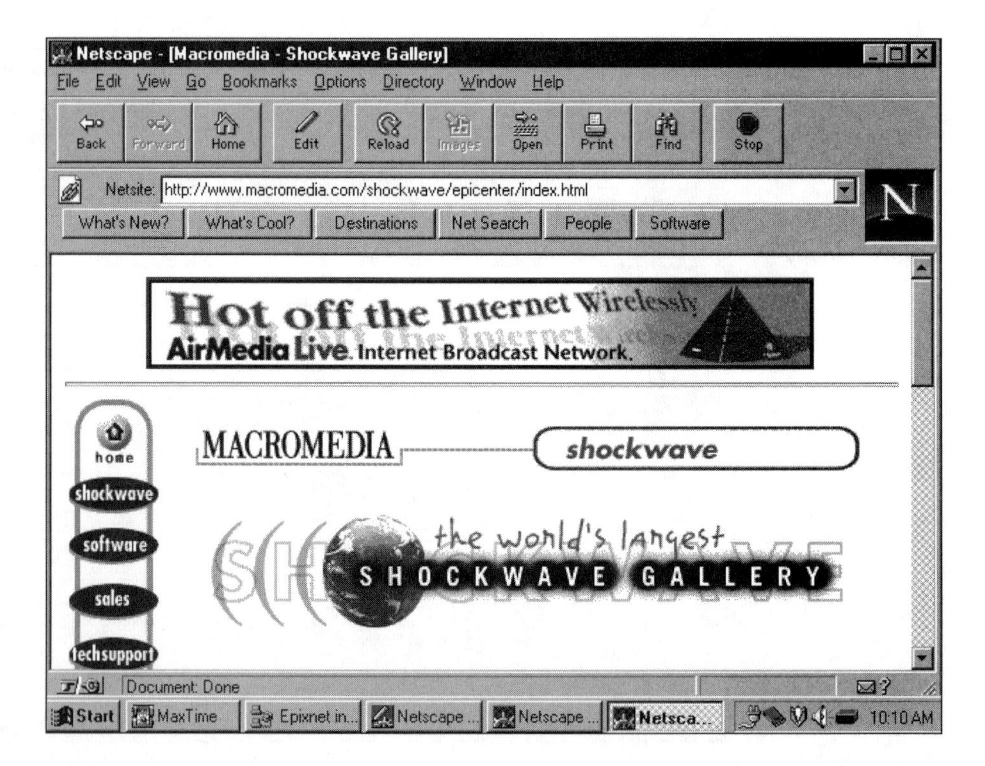

Figure 9-14. The Shockwave Gallery

Questions for Review

1. What is Adobe Acrobat? What are the characteristics of the PDF file format?
2. What are the advantages of using Acrobat files?
3. What are the steps and procedures for using Adobe Acrobat?
4. How is the Adobe Acrobat reader downloaded?
5. How is Adobe Acrobat configured for use as a helper application in Netscape?
6. What are the procedures for linking to a PDF file on the Web?
7. What are the parts of the Acrobat reader? What are the purposes of each part?
8. What are the functions of each button on the Acrobat reader toolbar?
9. How do you find PDF files on the Web?
10. What is ASAP WebShow?
11. What are the characteristics of ASAP WebShow?
12. What is Shockwave?
13. What are the advantages of using Shockwave?

Exercises

1. Download and install the Adobe Acrobat reader, if necessary (http://www.adobe.com/Acrobat). Configure it as a helper application for Netscape or other browser that you are using.
2. Use Netscape or another browser to download the Customer Trends: World Wide Web Publishing with Adobe Acrobat Software PDF file from the Adobe page at http://www.adobe.com/studio/spotlights/main.html#acrobat. You may want to print the six pages and then answer the following questions pertaining to the file.

   a. What is the title of the screenshot on page 1 of the document?
   b. Who is the author of the pull quote on page 2?
   c. What is the name of the newspaper from which screenshots on page 3 were taken?
   d. Who is the author of the pull quote on page 4?
   e. What are customers accessing the Dial-A-Book Download Bookstore able to do?

   f. Examine the caption on page 5 to determine the name of the CD the screenshot illustrates. What is the name of the company offering this catalog online?

3. Use Netscape or another browser to access Dial-A-Book Download Bookstore's page (http://www.psi.net:80/DownloadBookstore/index.html). List the titles and authors of three books that are being offered for download in PDF file format.

4. Use Netscape or another browser to access the U.S. Internal Revenue Service's tax forms page at http://www.irs.ustreas.gov/prod/forms_pubs/forms.html. Download the PDF version of Form 1040–U.S. Individual Income Tax form for last year. Save and print the form.

5. Use Netscape or another browser to link to the Internet (Timesfax) Edition of *The New York Times* (http://nytimesfax.com). Click on the Acrobat logo next to Click Here for TimesFax. Answer the following questions about today's issue.

   a. What are the titles of the lead stories on the front page?
   b. What are the titles of the lead stories on the international page?
   c. Who wrote the editorial on the commentary page?

6. Use Netscape or another browser to link to the "Guide to Silent Film Video Collection" in the Mitchell Multimedia Center at Northwestern University (http://www.library.nwu.edu/media/docs/). What are the titles of three videos in the film collection displayed in PDF file format?

7. Use Netscape or other browser to link to the Intertext Magazine (http://www.etext.org/Zines/InterText). Click on the link to the PDF version of the current issue of the magazine. Answer the following questions.

   a. What is the nature of this magazine?
   b. What is the editor's name?
   c. Explain the procedures for submitting articles to the magazine.

8. Use Netscape or another browser to link to Adobe's PDF Customer Spotlights at http://www.adobe.com/spotlights/main.html#acrobat. Click on two PDF files to download them. Print each one and write a paragraph for each file describing its contents.

9. Use Netscape or another browser in which ASAP WebShow has been installed as a plug-in or helper application to link to Software Publishing Corporation's Demo Center (http://www.spco.com/demo/wpwsdemo.htm). Link to several URLs that provide examples of how WebShow is being used by businesses today.

10. Write a paragraph for each of the following products and services describing how WebShow might be used on the Web.

    a. An insurance company needs a way to train its national salesforce.
    b. A new company would like to promote Window Glow, a new revolutionary window-cleaning product.
    c. The federal government would like to encourage people to purchase U.S. savings bonds.
    d. An Internet service provider needs a means for announcing its new nationwide Internet connections that will be available in three months.

11. Download and install the Shockwave plug-in in Netscape or another browser, following the steps provided in the chapter. The URL is http://www.macromedia.com/shockwave/download.

12. Make sure that the Shockwave plug-in has been installed in your browser. Link to the Shockwave Gallery at http://www.macromedia.com/shockwave/epicenter/index.html. Perform the following tasks and answer the questions.

    a. Click to link to the "Shocked Site of the Day & Week." What is the name of the shocked site for today? Link to the site. Describe how this site uses Shockwave.
    b. Click to link to the Corporate listing. What are the names of the companies that use Shockwave? Link to one of the sites. Describe how this site uses Shockwave.
    c. Click to link to the Education listing. What are the names of the education sites that use Shockwave? Link to one of the sites. Describe how this site uses Shockwave.
    d. Click to link to the Entertainment listing. What are the names of the entertainment sites that use Shockwave? Link to one of the sites. Describe how this site uses Shockwave.

13. Make sure that the Shockwave plug-in has been installed in your browser. Link to Marc's Shockwave Shop (http://mediaband.com/shockwave). This site provides examples of applications and resources that use Shockwave. Perform the following tasks and answer the questions.

    a. What are the different genres of Shockwave movies? Link to one or more of these sites.
    b. What are the different artists for which Shockwave movies exist?
    c. What other categories of Shockwave applications and resources are available at this Web site?

Case

Xavier University would like to provide students and faculty with easy access to computer documentation by using an online information network connected to the Internet. The university computing system consists of a series of file servers and several computer labs containing more than 130 Macintosh and 190 PC computers. It also provides students, faculty, and alumni who own modem-equipped computers with electronic access to university servers. The university maintains a homepage on the World Wide Web.

Write a paragraph discussing the advantages and disadvantages of using PDF files versus using HTML document files for the documentation. How should Xavier make the Acrobat reader available for users accessing information off campus?

CHAPTER

10

Virtual Reality on the Internet

This chapter describes virtual reality on the Internet and the types of free or inexpensive virtual reality browsers and viewers available for both the IBM and IBM PC–compatible and Macintosh computers. The steps and procedures for using Netscape Live3D will be explained.

What You Will Learn

- How virtual reality is defined
- Potential uses of virtual reality
- Using the virtual reality modeling language (VRML)
- Types of virtual reality browsers and viewers
- Steps and procedures for using Netscape Live3D

Virtual Reality Defined

If you've played one of the latest video games, you may have already experienced virtual reality, the newest multimedia tool to make the World Wide Web easier to use and closer to the types of three-dimensional (3-D) graphic presentations already featured in video games. The term **virtual reality** or **virtual world** may be defined as a created environment that uses a virtual reality modeling tool within which users operate and interact.

On the Web this concept is made possible by the **virtual reality modeling language (VRML)**, a language for describing multiparticipant interactive simulations in which virtual worlds are networked through the World Wide Web. VRML is a separate language altogether from the hypertext markup language (HTML) described in Chapter 3. VRML will someday provide opportunities for individuals to experience Web sites consisting of sophisticated three-dimensional sites very different from the one- or two-dimensional images that require users to initiate the interaction by pointing and clicking with a mouse.

The history of VRML can be traced back to the first World Wide Web conference held in Geneva, Switzerland in 1994. Conference attendees, many of whom were already involved with the development of three-dimensional visualization tools, agreed that there was a need to develop a common language that could deliver networked "scene descriptions," which they dubbed Virtual Reality Markup Language. Later the term *markup* was changed to modeling to better reflect the nature of the project's objectives.

Current and Potential Uses of Virtual Reality

VRML has several current as well as potential uses, including online conferencing, virtual storefronts, entertainment, and other applications. Let's look briefly at each of these applications.

ONLINE CONFERENCING

This application is closely aligned to the coverage of computer conferencing and videoconferencing in Chapters 7 and 8. Using VRML, it may be possible for private companies or other organizations to create a shared, virtual meeting space in which individuals meet to discuss corporate or organizational strategies and make decisions without leaving their offices.

VIRTUAL STOREFRONTS

There are already Web sites that offer products for sale and allow participants to purchase these items online. The next step to online shopping will be to incorporate VRML to allow individuals to explore departments or rooms in which merchandise is displayed. Once the problems of Web security are solved, we will probably see this application in a very short period of time.

ENTERTAINMENT

Interactive movies and virtual reality games are likely to appear on the Web in the not-too-distant future. There are already several sites that offer virtual reality–based entertainment on the Web. One of these is Virtual Vegas (http://www.virtualvegas.com).

OTHER APPLICATIONS

Other potential applications for VRML include scientific and medical research and architecture. Several universities have already begun to use VRML to develop three-dimensional models for cell membranes and atomic orbitals. It has also been used in projects to define astronomical objects such as galaxies and black holes. Architecture is a very likely area for VRML, which makes it possible to erect and tear down proposed structures so that builders will have a clearer idea of construction and design requirements.

Using the Virtual Reality Modeling Language (VRML)

The virtual reality modeling language (VRML) can be created in the following three ways: (1) by creating and editing a VRML text file by hand; (2) by using a conversion program to convert an existing non-VRML file to VRML; and (3) by using an authoring package to create models and position them within a virtual reality world. Any of these methods will produce a VRML file that usually has a "WRL" extension. It is very possible that you may need to slightly modify VRML files created with an authoring tool or converter. This is because VRML is still a very new technology.

You may want to familiarize yourself with the basic VRML syntax and concepts. The following is a file that provides an example of VRML coding. Exercise 4 in this chapter provides practice in working with this file.

```
#VRML V1.0 ascii
#by Michael J. Donahue, (mediastorm@pacificnet.net)
#http://pacificnet.net/~mediastorm
#This is a hack on John Walker's(kelvin@fourmilab.ch)code
#Earthview at http://www.fourmilab.ch/earthview/vplanet.html
#He really did the work and Earthview is worth checking out!
Separator {
    ShapeHints {
        vertexOrdering COUNTERCLOCKWISE
        shapeType  SOLID
        faceType   CONVEX
      }
    Material {
        diffuseColor 1 1 1
```

Continued on next page

Continued from previous page

```
        }
      Texture2 { filename "http://www.fourmilab.ch/earthview/Current-clouds.gif" }
      Sphere { radius .75 }
   }
```

THE EMBED TAG

After the VRML world has been created, it may be embedded within a HTML document by using the <EMBED> tag. This is similar to using the <IMG> tag that is used to place a two-dimensional image in a HTML document (see Chapter 3). The following example embeds a VRML file called example.wrl into a HTML document.

```
<EMBED SRC="example.wrl" WIDTH=128 HEIGHT=128 BORDER=0 ALIGN=middle>
```

VRML ELEMENTS

The speed and effectiveness of VRML worlds are affected by a number of factors. The developers of Netscape's Live3D viewer (see later) list the six elements that will affect a Web page using VRML: polygons, textures, instancing, level of detail, inlines, and compression.

In a VRML world shapes are made of *polygons*—the more complex a shape, the more polygons are required. A cube, for example, is typically composed of just twelve polygons, since each side is made of two triangles. In contrast, a simple sphere requires more than 200 triangular polygons. As more objects are added to a world, the polygon count for that world increases. Each time a user's viewpoint changes in the VRML world, the browser has to redraw the scene. The more polygons the world contains, the longer the redraws take.

Thus low polygon counts are one way to increase navigation speed. VRML allows the textures to be mapped onto shapes, and textures used in a VRML world may increase its size considerably, affecting both download and redraw times. Therefore, if textures are used, small textures are desirable as one way to keep download times low and navigation speed high. Textures used in VRML worlds will also require fewer client resources if they use fewer colors.

Once they are defined, objects may be reused in a VRML world. This feature can help to keep a world's file size small. This technique of reusing objects is called **instancing**. Although there are some limitations to instancing, it can make VRML code easier to write and maintain and VRML worlds easier to download.

In the real world, more details become visible as you get closer to an object. The Level of Detail node (LOD) makes this possible in VRML worlds. The LOD node determines which objects will be visible within defined ranges of coordinates within the VRML scene. This feature permits special effects and realistic simulations.

Other world files may be "pulled into" a world to help create a VRML scene. Used this way, these files are called *inlines*. The WWW Inline node is used to refer to a world file to be included and, optionally, to display a bounding box to show the user where the object, or objects, will be positioned before they are rendered.

The larger the VRML world file, the longer it takes to download. World files may be compressed using utilities such as GZIP. If a VRML browser recognizes the file type, it can automatically parse the compressed file to display the VRML world.

Types of Virtual Reality Browsers and Viewers

Not all the standard HTML browsers discussed in Chapter 2 can be used to view VRML files. To view a VRML file on the Internet, a VRML browser or viewer is needed. Some can serve as standalone browsers, whereas others are helper applications or plug-in modules that function with Netscape and other HTML browsers.

Plug-in modules (also known simply as **plug-ins**) are software programs that extend the capabilities of Netscape and other HTML browsers. A plug-in is installed on your hard disk using instructions that come with it. After installation, the HTML browser uses the plug-in's capabilities like other built-in browser features.

Plug-ins can have one of three modes of operation: embedded, full-screen, or hidden. An *embedded* plug-in is a part of a larger HTML document, visible as a rectangular frame within a page (embedded plug-ins are specified in HTML with the EMBED tag). A *full-screen* plug-in is a self-contained viewer, completely filling the content area of a Netscape window. A *hidden* plug-in runs in the background.

The VRML browsers and viewers now available are written to work with the Windows 3.1, Windows NT, Windows 95, SGI, Sun, UNIX, and Macintosh platforms. Most of the Windows VRML viewers are written for Windows NT or Windows 95. Only a few VRML browsers and viewers are written for the Macintosh computer. Most VRML browsers and viewers are capable of reading standard VRML files with the WRL extension. The following is a brief description of some of the many VRML browsers and viewers currently available.

VIRTUS VOYAGER

Virtus Voyager may be used as a standalone product or as a helper application for Netscape or Mosaic browsers. Voyager reads VMRL files but can be expanded to import Virtus's own proprietary VMDL file format, which is efficient for larger files. Voyager is available for Windows 95 and Macintosh/Power Macintosh platforms. The browser can be downloaded at http://www.virtus.com/voyager.html. Figure 10-1 shows the Voyager display of a VMRL file in the Macintosh version.

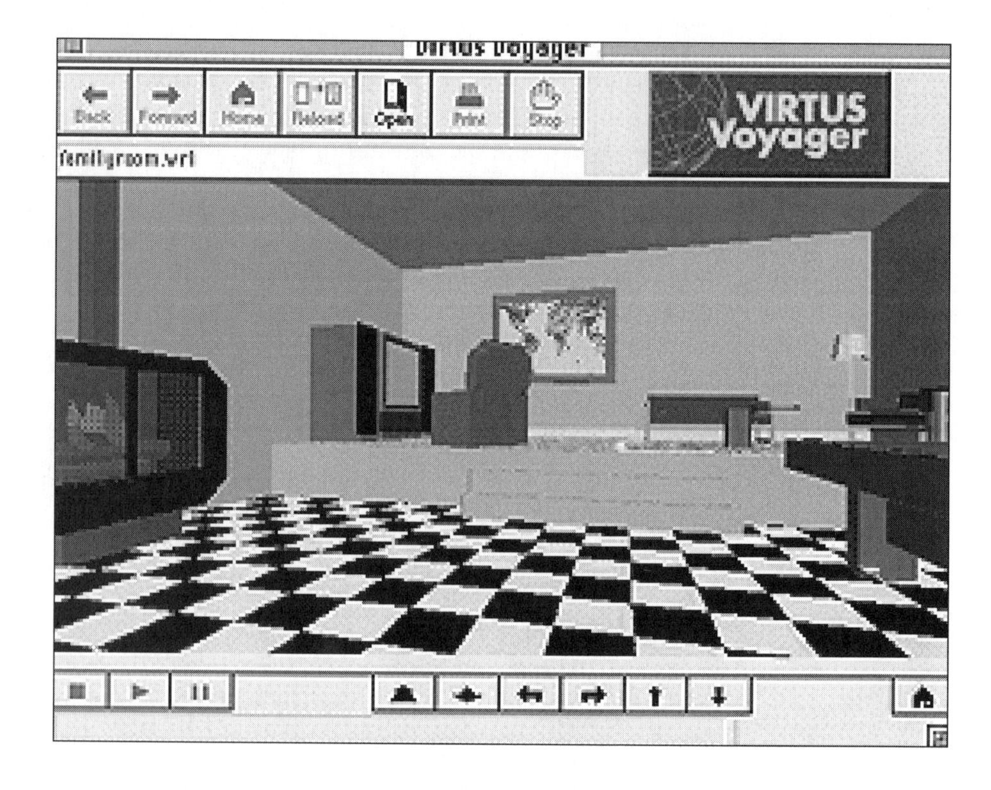

Figure 10-1. The Macintosh Version of Virtus Voyager

VR SCOUT VIEWER

VR Scout is a viewer rather than a standalone browser that supports all VRML specifications including Inlines, LOD, Anchors, textures, and the like. It also supports gzipped files, zipped files, GIF/JPEG/BMP textures, DDE, internal HTML if no DDE browser is available, and threading on Windows 95 and Windows NT. The following platforms are supported: Windows 3.1, Windows NT, and Windows 95 (works with Netscape Navigator 2.0 or later). The external viewer works on Windows 3.1, Windows 95, and Windows NT 3.51. The external viewer is made for Web browsers that require an external "helper" application, such as Mosaic, Netscape (16-bit versions), and other browsers. VR Scout is available for download at http://www.chaco.com/vrscout/#about. Figure 10-2 shows the VR Scout being used to view a virtual reality file in Netscape as a plug-in.

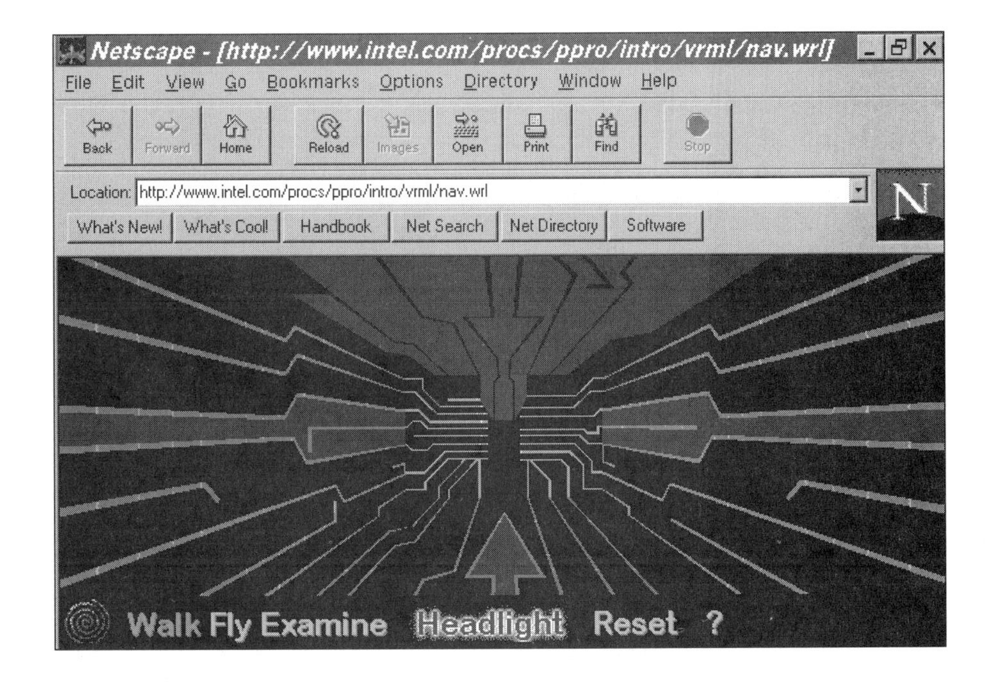

Figure 10-2. VR Scout Viewer in Netscape

MICROSOFT INTERNET EXPLORER VRML ADD-IN

Microsoft's Internet Explorer VRML add-in is designed for the Microsoft Explorer HTML browser, Version 2.0 for Windows 95 (see Chapter 2). This VRML viewer runs as an add-in to the Explorer browser and becomes a part of the browser after it is installed. In comparison with other VRML browsers and viewers, it offers high-speed performance. The Microsoft Internet Explorer VRML Add-In can be obtained at http://www.microsoft.com/ie/ie3/vrml.htm. The Microsoft Explorer browser must be installed before the Add-In is installed. Figure 10-3 shows a VRML file using Microsoft Explorer and the VRML Add-In. (Note: The Microsoft Internet Explorer browser, version 3.0 and above, has the ability to view VRML files as a built-in feature.)

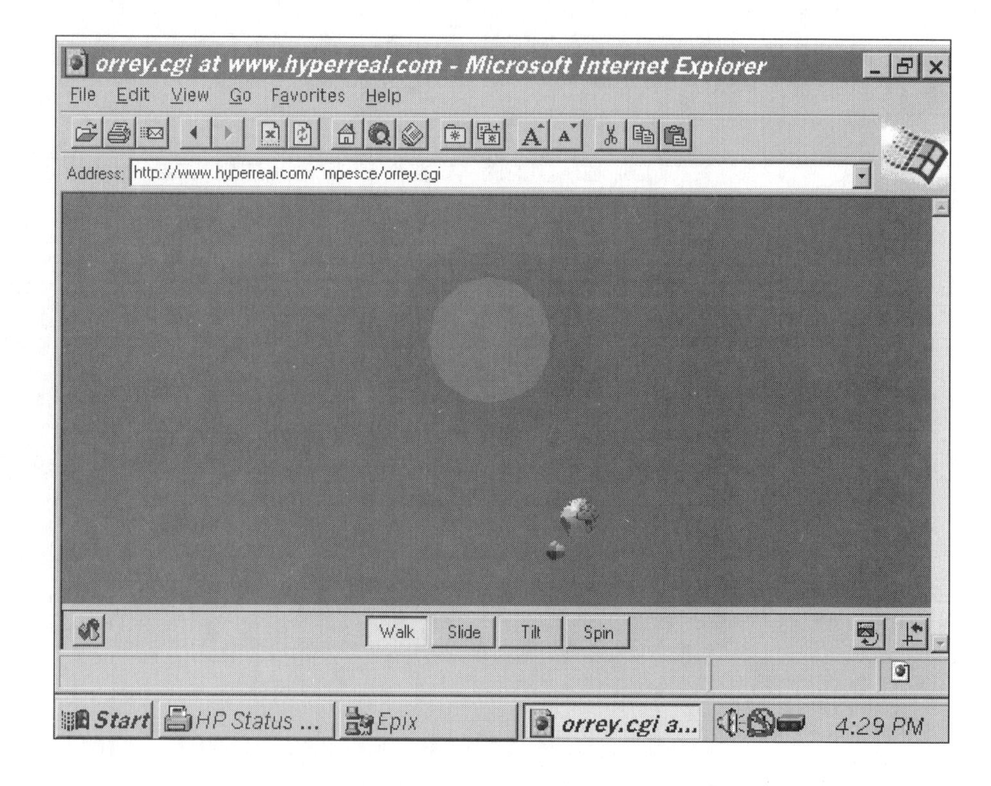

Figure 10-3. The Microsoft Internet Explorer with the VRML Add-In

NETSCAPE LIVE3D

The Netscape Live3D viewer is a Netscape Navigator add-on that makes it possible to experience distributed, interactive virtual reality files on the Internet. Live3D also extends Netscape's Java, JavaScript, and plug-in interfaces, making it easy to develop distributed three-dimensional applications and to experience text, images, animation, sound, music, and video in a 3-D realm.

Live3D possesses high-performance VRML viewing at maximum speed with adaptive rendering, hardware acceleration, background processing, and GZIP data compression. The real worlds can be full of lifelike behaviors with full integration of audio and video. Navigation with Live3D takes place by letting you walk, fly, or point. Collision detection and selectable camera viewpoints as well as optional gravity add flexibility and realism to navigation.

The Walk command allows you to navigate to a specific place in the virtual world. To use the Walk command, you first click on "Walk" on the command line. Dragging the mouse arrow in the direction you would like to walk in the virtual world will move you in that direction. The Point command, when selected, allows you to move to a specific object in the virtual world. The Look command allows you to link to another world view or a HTML page linked to a virtual world object.

Live3D is based on the proposed Moving Worlds VRML 2.0 specifications that have been endorsed by many companies and recently submitted to the VRML Architecture Group. Live3D comes packaged with Netscape 3.0 as a plug-in module and is automatically installed as a part of the installation procedure. If you installed Netscape 3.0 without Live3D, it can be installed later as an plug-in module. When you encounter a .WRL file on the Web while using Netscape, the Live3D viewer is automatically activated. Figure 10-4 shows the Live3D viewer within Netscape with a linked VRML file.

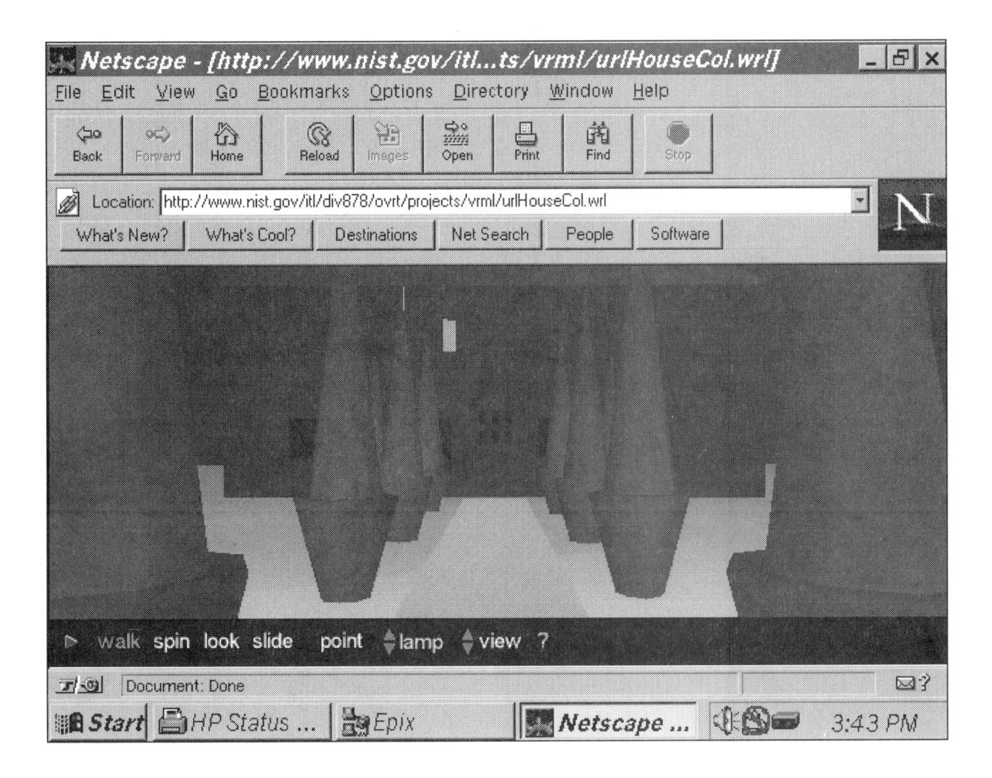

Figure 10-4. The Netscape Live3D Viewer

Steps and Procedures for Using Netscape Live3D

Here's how to install and use Netscape Live3D.

DOWNLOADING AND INSTALLING NETSCAPE LIVE3D

If you are using a version 3.0 of the Netscape browser, the Live3D viewer will need to be downloaded and installed as a plug-in module if you didn't request it to be part of the initial installation. If Live3D has already been installed, skip these steps. Follow these steps to download the viewer.

1. Use Netscape to open the following Web page: http://home.netscape.com/comprod/products/navigator/live3d/download_live3d.html. You can also use the address ftp://ftp.netscape.com/pub/live3d and find the correct version of Live3D to download.
2. Run the downloaded exe and a number of files will be extracted. Run the setup.exe file and follow the onscreen instructions. You will be asked where you would like the Live3d files to be installed on your hard drive.

USING NETSCAPE LIVE3D

After Live3D has been installed, it is an easy procedure to use the viewer. When you encounter a VRML file, which usually has a WRL extension, Live3D will automatically allow you to view the file. Follow these steps to practice using Netscape Live3D.

1. Load Netscape 3.0 with the Live3D viewer installed.
2. Open the following URL: http://www.zdnet.com/zdi/vrml/content/vrmlsite/outside.wrl. This should open Ziff-Davis's virtual world. It may take some time to load in all of the parts of the world (see Figure 10-5).

   You will notice that the screen shows seven commands that are a part of Live3d: Walk, Spin, Look, Slide, Point, Lamp, and View. The Walk command is used when you want to walk to a specific place in the virtual world and is the default. You click on the other commands to activate each of them.
3. Move the mouse arrow on various places of Ziff-Davis's World and an identifier will pop up. You can examine all of the parts now, but do *not* click the mouse. Move the mouse arrow on the door of the terminal and the identifier Click Here to Enter Terminal Reality and click the mouse. The portion of the world in Figure 10-6 showing a man on an elevator should come into view. Notice that each of the items on the front of the elevator represents different magazines that Ziff-Davis publishes.

Figure 10-5. The Initial Ziff-Davis World View with Live3D

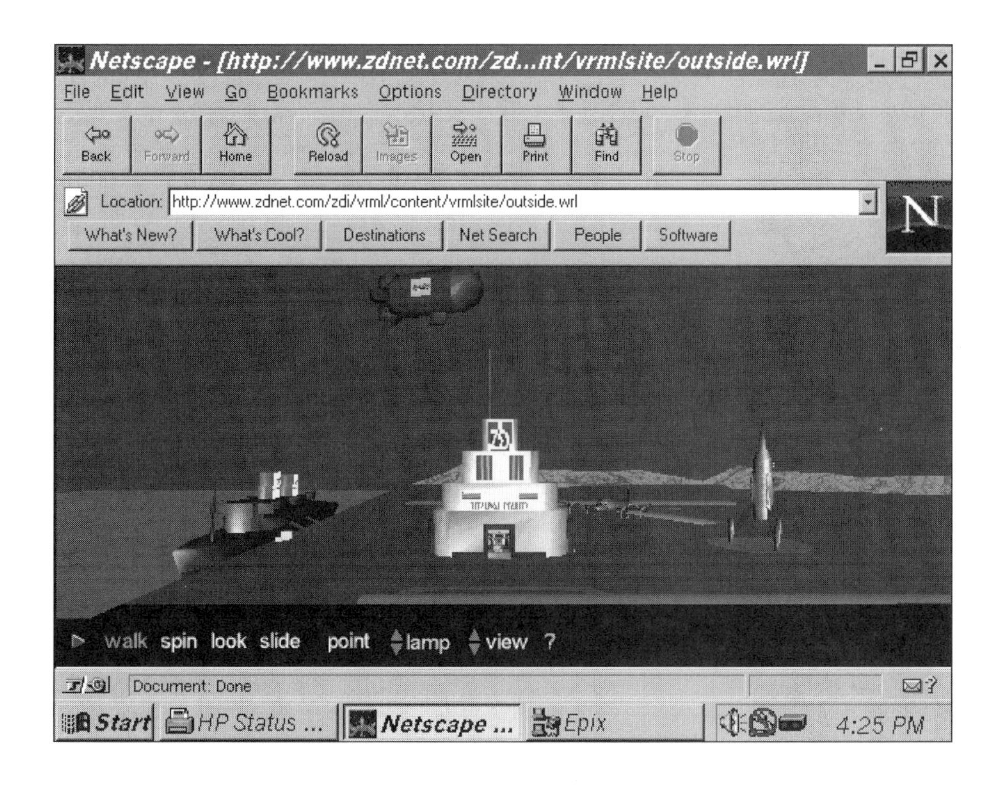

Figure 10-6. The Second Ziff-Davis World View with Live3D

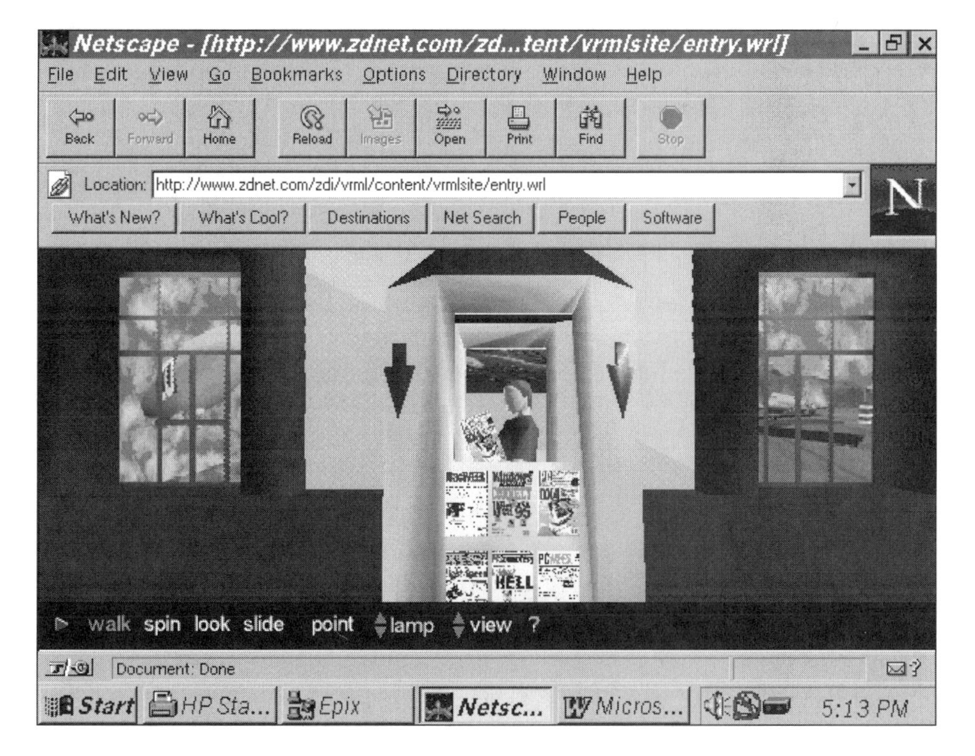

4. Click on the window at the right. The identifier should be Fly the ZD Airplane. This should load the main world view again. To get a better view of the airplane, we will need to spin the view so that we will be looking down on the runway.
5. Once the main world view is loaded again, click on Spin on the command line to select it. Place the mouse arrow at the bottom of the runway. Hold down the mouse button and drag the arrow down toward the bottom of the browser screen. This should give you a bird's-eye view, showing the airplane to the right of the terminal building (see Figure 10-7).
6. Click on the airplane and the caption "Fly the ZD Airplane" should appear.
7. Experiment with the Spin command to obtain different views.

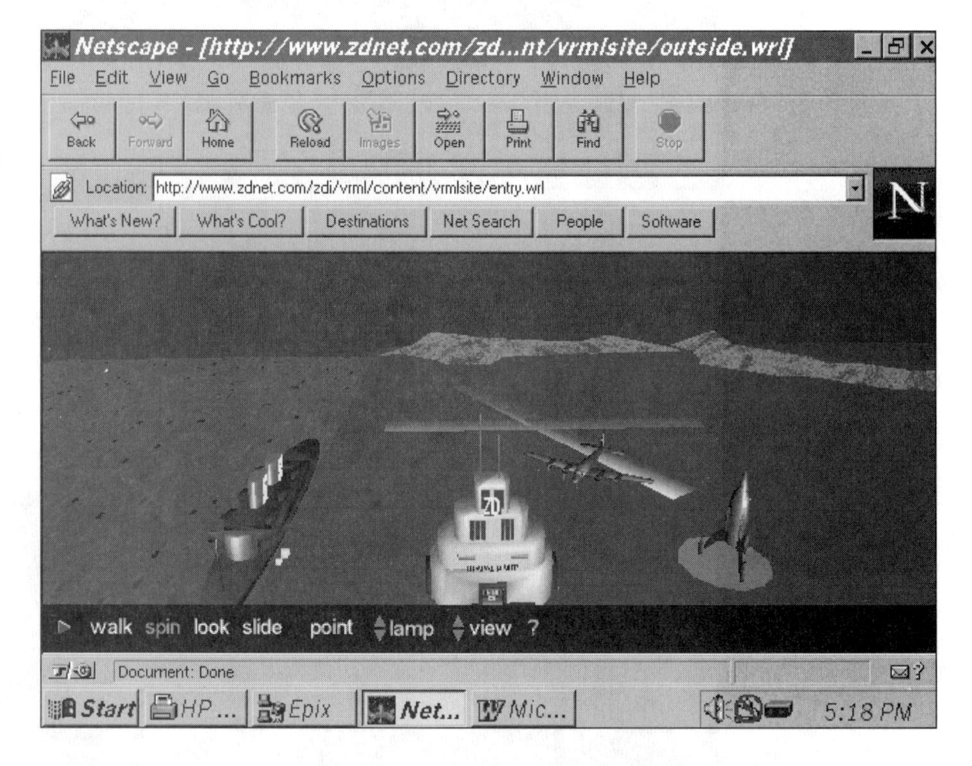

Figure 10-7. The Initial Ziff-David World View After Using the Spin Command with Live3d

Questions for Review

1. What is virtual reality or a virtual world?
2. What is the purpose of the virtual reality modeling language (VRML)?
3. What are the potential uses for virtual reality?
4. What is the purpose of the <EMBED> command in HTML?
5. What are the six VRML elements that will affect a Web page using VRML? Describe each of these elements.
6. What is instancing? What are inlines?
7. What is the Level of Detail node (LOD)?
8. Name the types of VRML browsers and viewers.
9. What is a plug-in module? How does it differ from a helper application?
10. Explain the steps and procedures for using Netscape Live3D.

Exercises

1. Download and install one of the VRML browsers, plug-ins, or viewers if necessary.
2. Link to one of the following Web sites that contain virtual worlds or links to worlds and resources. Load one or more of the virtual worlds using Netscape or some other browser that supports VRML.

   http://home.netscape.com/comprod/products/navigator/live3d/examples/examples/examples.html
   http://www.nist.gov/itl/div878/ovrt/projects/vrml/vrmlfiles.html
   http://www.autonomy.com/virtual.htm
   http://www.clark.net/theme/proteinman/
   http://home.netscape.com/comprod/products/navigator/live3d/cool_worlds_fs.html

3. Use Netscape and Live3D or some other browser that supports VRML to access the Ziff-Davis's Virtual World at http://www.zdnet.com/zdi/vrml/content/vrmlsite/outside.wrl. Answer the following questions about this world.

   a. What is the caption for the blimp?
   b. What is the caption for the rocket?
   c. What is the caption for the ship?

d. Click on the terminal door in the Mail World View and retrieve the world view containing the elevator. What are the titles of the six magazines on the front of the elevator?
e. Click on the blimp on the Main World View to go to Snailworks. What are the names of the four locations?

4. Use Netscape with Live3D to access pacific.net's earth view at http://www.pacificnet.net/~mediastorm/earthrt.wrl. This shows the current cloud cover. (See the VRML code for this page that was illustrated in this chapter.) Answer the following questions.

a. Use the Spin command to spin the earth so that North America is in view. Are there any clouds covering the continent?
b. Use the Spin command to tilt the earth so that you can see the North Pole. Is there any cloud cover over the North Pole?

5. Use Netscape with Live3D to experience how Duke's Diner from Market Central use Live3D and RealAudio at http://www.marketcentral.com/vrml/duke.wrz. You will be clicking on various parts of the world to try to find Duke, the owner of the diner, using RealAudio audio clips. (Note: You must have RealAudio installed to be able to hear the audio clips. See Chapter 5 for information on RealAudio.)

6. Use Netscape and Live3D to access http://www.virtpark.com/theme/worlds/ab2.wrl. Use the Walk command to walk into the building. Find the sun and use the Look command to click on it and link on sun.html. Answer the following questions about the sun.

a. What type of star is the sun?
b. What is the sun's diameter?
c. What is the core temperature of the sun?
d. What is the surface temperature of the sun?

7. Use Netscape and Live3D to access a virtual Italian café at http://www.construct.net/projects/planetitaly/Spazio/VRML/siena.wrl. Use the Walk, Look, and Point commands to explore this virtual world.

APPENDIX

A Downloading Files

This appendix will provide instructions for downloading files from a remote server connected to the Internet to your computer using both IBM or IBM PC–compatible and Macintosh computers.

On the IBM or IBM PC–compatible and Macintosh computers there are four methods of downloading files from a server connected to the Internet. This appendix will provide examples using each of these methods.

What You Will Learn

- File transfer protocol (FTP) from the UNIX prompt (IBM only)
- File transfer protocol (FTP) with a Windows utility such as WinFTP (IBM only)
- File transfer protocol (FTP) uniform resource locator (URL) (ftp://) with a Web browser such as Netscape (both IBM and Macintosh)
- Fetch utility (Macintosh only)

As an example, follow the steps for each method below to download the text file netiquette.txt from ftp.sunet.se in the directory /pub/Internet-documents/doc /netiquette. If you are unable to connect to the ftp.sunet.se server, choose one of the following servers and paths:

Server: thedon.cac.psu.edu
Path: /pub/people/dlp/TRDEV-L

Server: hasle.oslonett.no
Path: /gopher

Using FTP from the UNIX Prompt (IBM only)

If you can access terminal emulation on your computer, you may be able to use this method of downloading files. Follow these steps for downloading the netiquette.txt file. (Note: UNIX is case sensitive. You must enter the characters exactly as they appear.)

1. Log in to your server in the usual manner.
2. At the UNIX prompt, type ftpftp.sunet.se and press Enter.
3. Key anonymous for the Name and press Enter.
4. Key your electronic mail address for the password and press Enter.
5. Key cd /pub/Internet-documents/doc/netiquette and press Enter.
6. Key ascii and press Enter. Note: If you were downloading a binary file (with exe, zip, or Z as the extension), you would type bin instead.
7. Key get netiquette.txt and press Enter.
8. After the downloading is completed, key quit to exit ftp.
9. At the UNIX prompt, key ls and the netiquette.txt file should be listed.

The file is downloaded to your local server. Ask your instructor or network administrator how to use ftp or another utility to download the file from the server to a floppy disk on your local computer. This file can be retrieved and printed using any word processing program.

Using FTP with a Windows FTP Program (IBM Only)

Another excellent way to download files from a server connected to the Internet is with a Windows ftp utility program such as WinFTP or WS_FTP. Follow these steps for downloading the netiquette.txt file. (Note: If a FTP utility program is already installed on your computer, skip to step 4.)

1. Download the WS_FTP program (WS_ftp.exe, WS_FTP.ZIP, or WS_ftp32.zip files) by entering the following URL using the UNIX prompt method or Netscape (see the next section): http://www.ipswitch.com/pd_wsftp.htm/.

   Server: ftp1.ipswitch.com
   Path: /pub/win3/ or /pub/win32/for Windows 95
   URL: ftp://ftp1.ipswitch.com/pub/pub/win3/ or
   ftp://ftp1.ipswitch.com/pub/win32/

   Server: fcs280s.ncifcrf.gov
   Path: /pub/net.tools/windows/ws_ftp/
   URL: ftp://fcs280.ncifcrf.gov/pub/net.tools/windows/ws_ftp

   Note: There are several versions of WS_FTP. Choose the WS_FTP Limited Edition if you have a choice.

2. Use a ZIP utility program (such as pkunzip.exe or WINZIP) to unzip the zipped file, if necessary. See your instructor or network administrator for instructions on how to unzip the file.

3. Set up the WS_FTP utility program on your computer. See your instructor or network administrator for instructions on how to set up this utility program.

4. Click on the program icon to execute the program.

5. When you first execute the program, the Session Profile box will be displayed (see Figure A-1).
   Click on the New button and key each of the following:

   Config: Sunet as the profile name; press the Tab key four times.
   Host: ftp.sunet.se press the Tab key twice.
   Userid: anonymous (or click on the Anonymous Log in box); press the Tab key.
   Password: (key your complete e-mail address); press the Tab key twice.
   Initial Dir: /pub/Internet-documents/doc/netiquette

6. Click on the OK button. The main program window should appear (see Figure A-2). (Note: If you fail to connect to this server, use one of the other servers listed on page 224.)

7. Place a floppy disk in drive A or B. Click on the ChgDir button and key a:/ or b:/. Your local files should appear on the left side of the main window.

8. Click to select the netiquette.txt file in the Remote host info window on the right side of the main window.

9. Click to select ASCII.

10. Click on the left arrow to download the file to the A or B floppy disks.

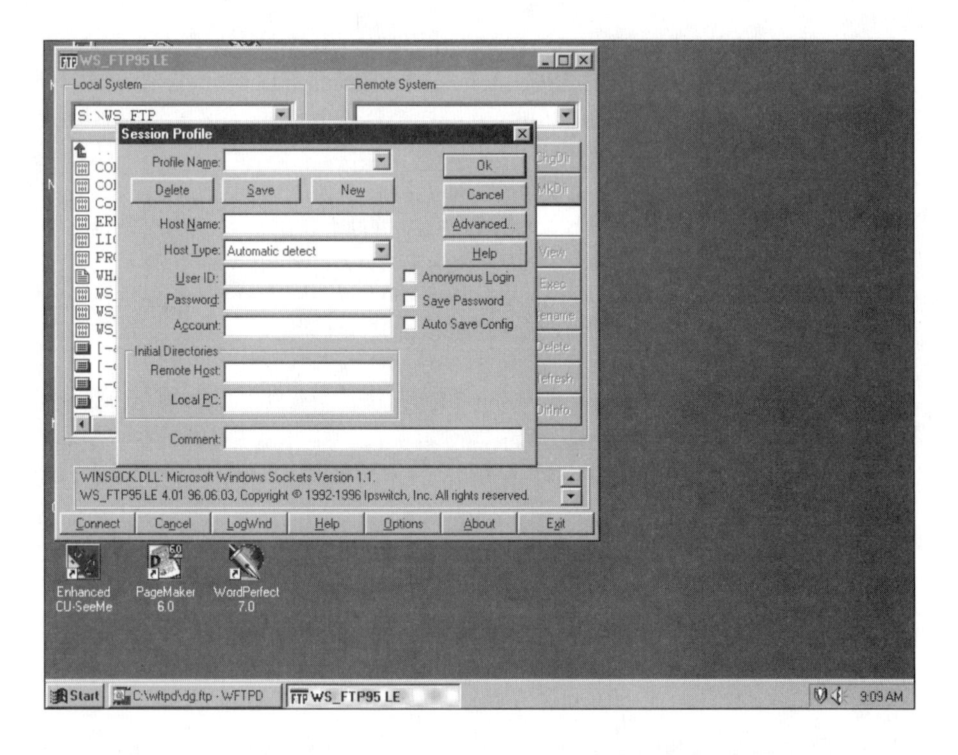

Figure A-1.
The WS_FTP Host Dialog Box

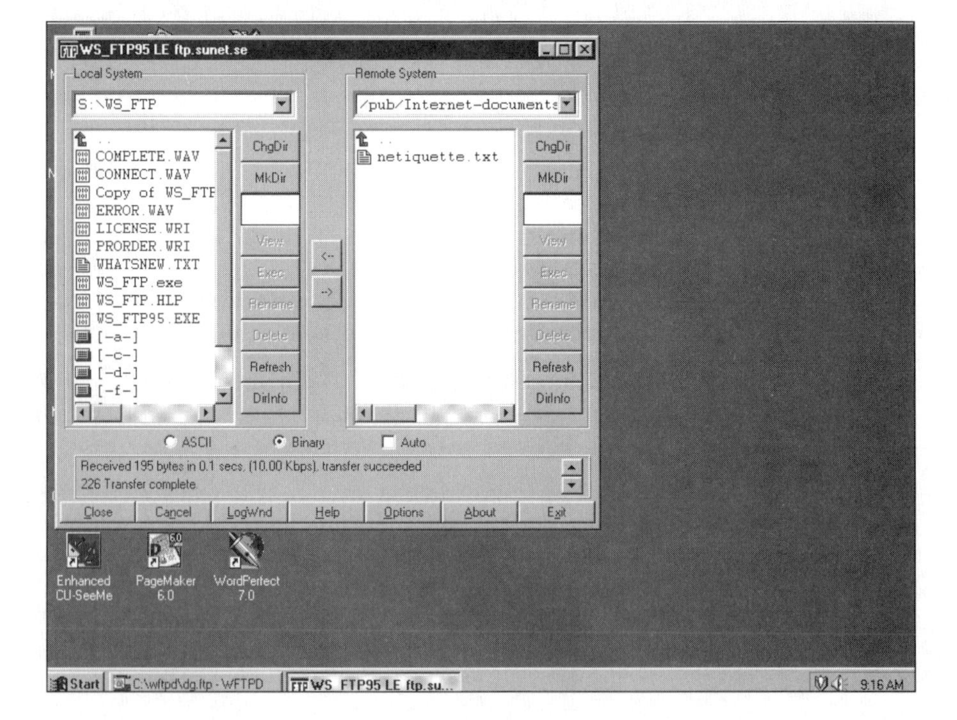

Figure A-2.
The Main WS_FTP Program Window

The netiquette.txt file should now be on your floppy disk. This file can be retrieved and printed using any word processing program.

Using a FTP URL with Netscape (Both IBM/IBM PC and Macintosh)

The third method of downloading a file from a server connected to the Internet to your local computer is to use Netscape or some other browser program. Click on the Open button on the toolbar (or choose Open Location on the File menu). Key in the URL listed here to connect to the specific server and to display the netiquette.txt file.

ftp://ftp.sunet.se/pub/Internet-documents/doc/netiquette/netiquette.txt

You can either save this file (File/Save As...), mail it to yourself (File/Mail Document...), or print it by clicking on the Print button on the toolbar or by choosing File/Print.

Fetch Utility (Macintosh Only)

If you are using a Macintosh computer, you can use the Fetch utility to transfer the file netiquette.txt to your local computer. Follow these steps.

1. Place a disk in the drive. Format the disk if necessary.
2. Access the Fetch utility. Figure A-3 shows a Fetch window. Your screen may display different information.
3. Key the following information.

   Host: ftp.sunet.se
   User ID: anonymous
   Password: (key your complete e-mail address)
   Directory: /pub/Internet-documents/doc/netiquette

   If you are not able to connect to this server, choose one of the other servers and paths listed on page 224.
4. Click on the OK button and save the file on your floppy disk.

Figure A-3.
The Macintosh Fetch Utility Window

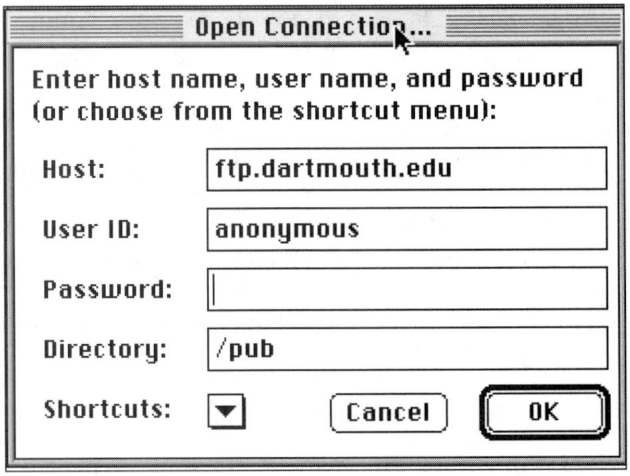

Open Connection...
Enter host name, user name, and password (or choose from the shortcut menu):
Host: ftp.dartmouth.edu
User ID: anonymous
Password:
Directory: /pub
Shortcuts:
Cancel
OK

APPENDIX

B

Additional Multimedia Web Sites

An increasing number of Web sites are utilizing multimedia. The following represents a sample of Web sites that use graphics, sound, video, and animation.

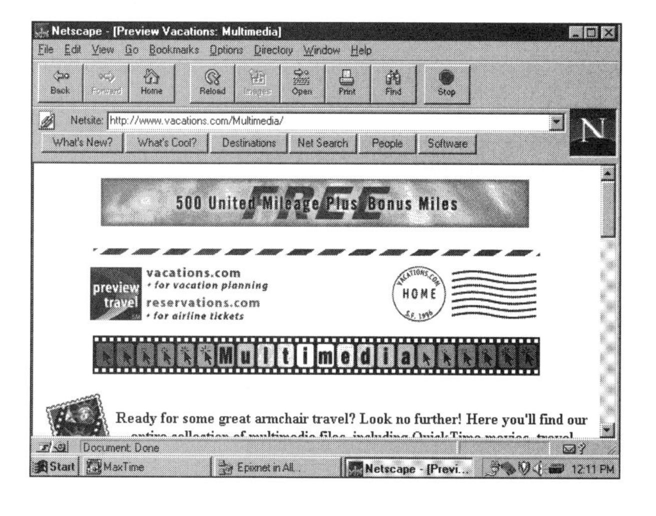

Preview Vacations
http://www.vacations.com/Multimedia

This Web site contains a collection of multimedia files, including QuickTime movies, images, QuickTime VR virtual tours, and ambient sounds that you can use to plan your vacation.

John Philip Sousa Home Page
http://www.dws.org/sousa/

The complete sound files of many of Sousa's works are accessible through this Web site. Included are over 60 sound clips, as well as video files of bands performing Sousa marches and graphics files of the composer.

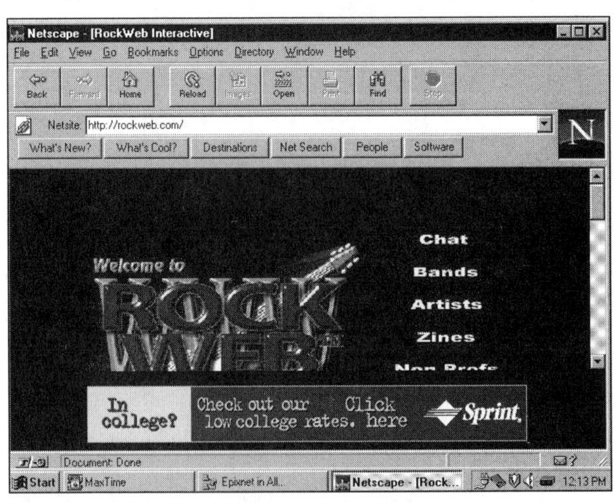

Rockweb Interactive
http://rockweb.com

This site contains video clips and sound files of popular rock bands.

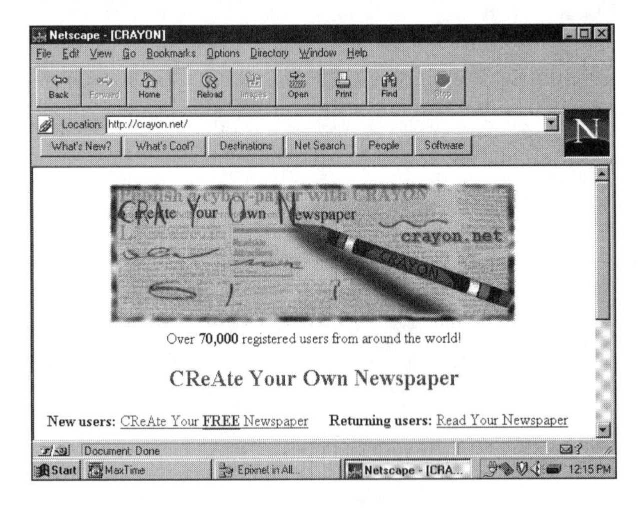

CRAYON
http://crayon.net

CRAYON is a tool for managing news sources on the Internet and the World Wide Web. It uses a simple analogy that everyone can understand—a newspaper to organize periodical information. The result is a customized news page with daily information tailored to your interests.

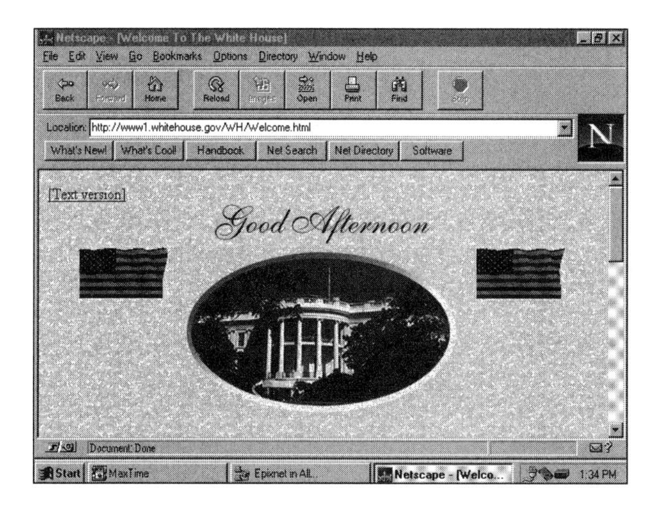

White House Web Page
http://www.whitehouse.gov

The White House Web Page contains information about the White House and the federal government. The current version contains two waving American flags using Java applets and a changing photograph of the White House. Information is provided about the current administration and White House history and tours. Sound files of important speeches made by the president and a White House section specifically designed for children are included.

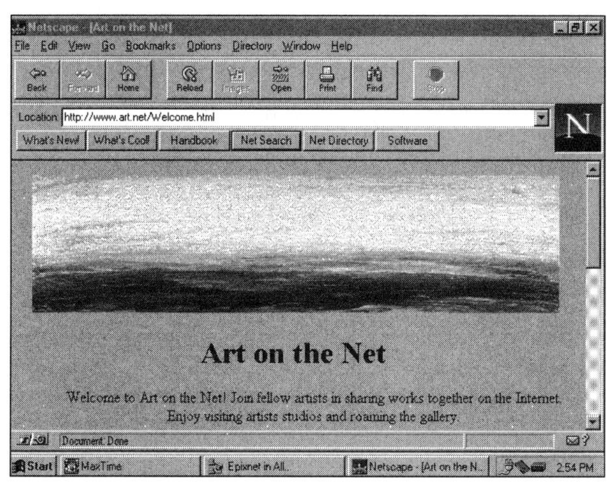

Art on the Net
http://www.art.net

This Web site is an online gallery featuring the work of international artists, from painters to poets and digital creators. You can wander the rooms, download artist sound clips, and peruse an events calendar.

Playbill On-Line
http://www.playbill.com
or
http://wheat.symgrp.com/playbill/

This Web site provides information about Broadway shows. The multimedia center section offers pictures, sound, and video clips of currently playing shows. Pictures include playbill covers, billboards, and production photos.

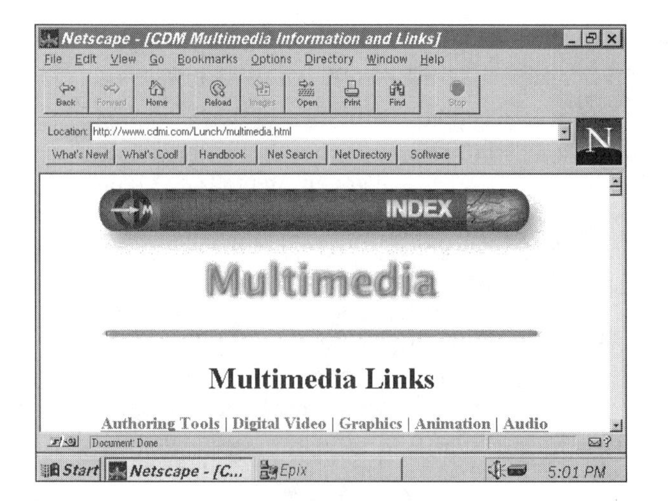

Multimedia Links
http://www.cdmi.com/Lunch/multi-media.html

This site provides links to authoring tools, digital video, graphics, animation, audio, business/legal, other multimedia indices, periodicals, and newsgroups.

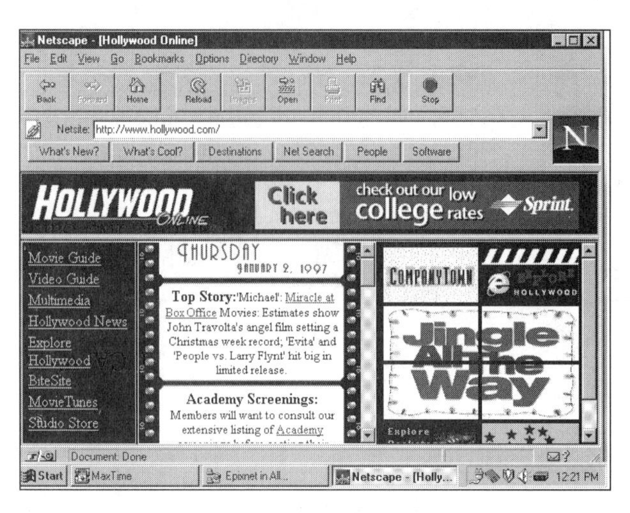

Hollywood Online
http://www.hollywood.com

This Web site helps you to locate films that are playing in theaters nationally and download photos, sound, video, and trailers.

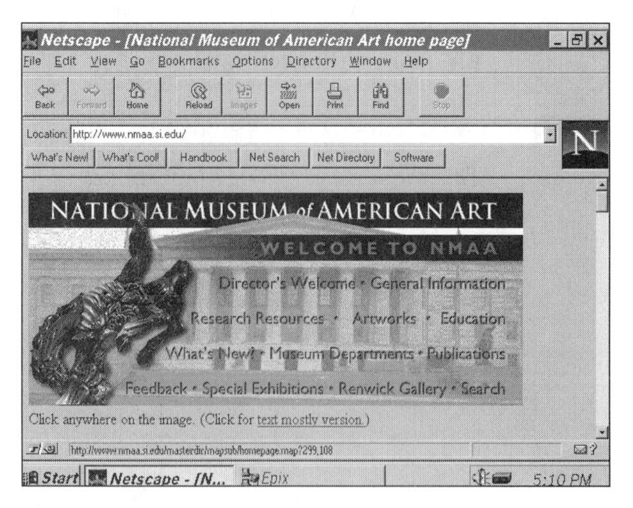

National Museum of American Art
http://www.nmaa.si.edu

This site represents the Smithsonian Institution branch that includes artwork from the permanent collection. Quicktime movies and RealAudio presentations are offered.

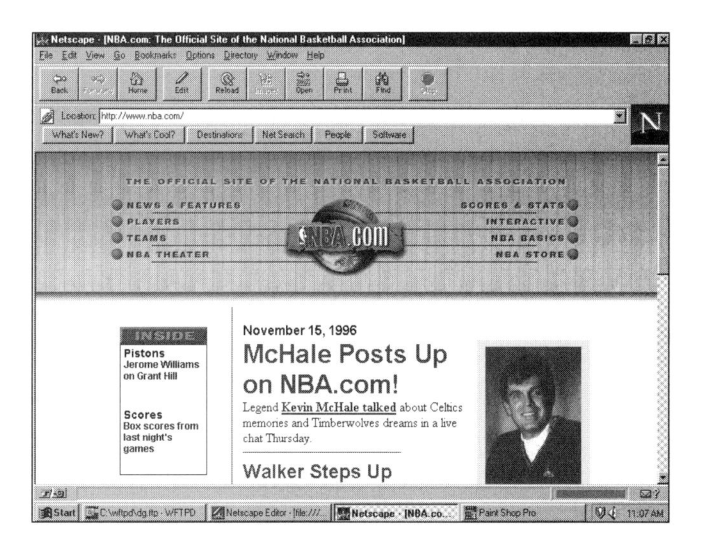

National Basketball Association
http://www.nba.com

This Web site provides highlights from the latest games, player interviews, and related features that you can see and hear.

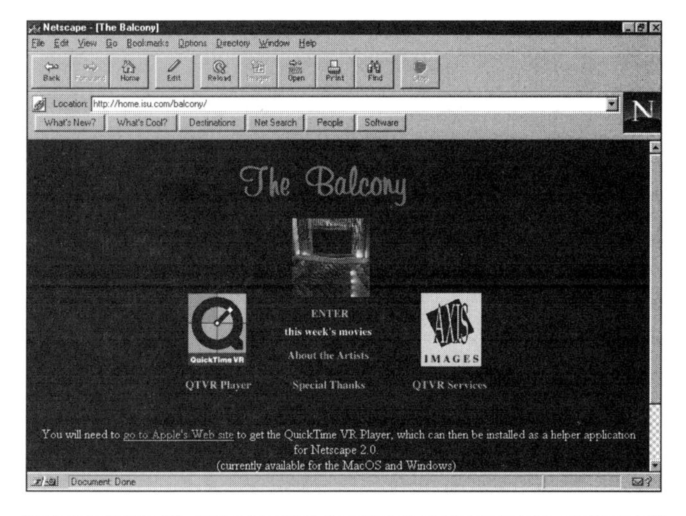

The Balcony...
http://netproductions.com/balcony

You are able to download the latest information relating to movie production recorded in Quicktime files with this site.

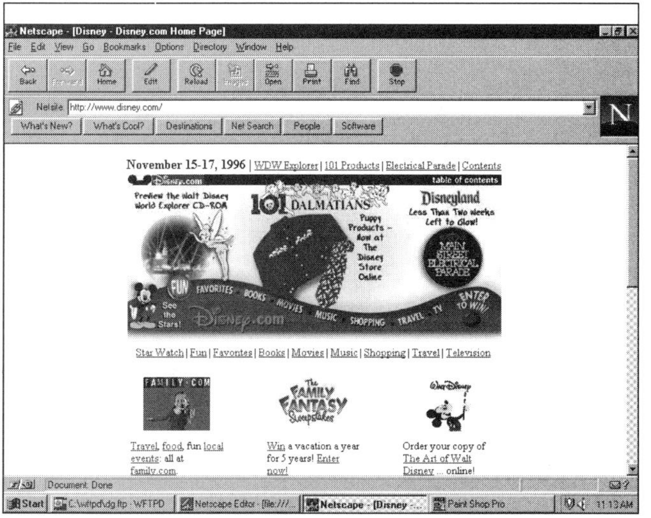

Disney
http://www.disney.com

This site provides information about the Disney Company, including current news about movies, music, television, travel, and tours. In Sights & Sounds, photo, sound, and video files may be downloaded.

GLOSSARY

Adobe Acrobat A document format that includes capabilities for retaining original page layout information, fonts, graphics, etc.

animation A multimedia element incorporating line art that creates the illusion of movement in a program.

anonymous ftp In systems connected to the Internet, the use of a file transfer program to contact a distant computer system to which you have no access rights, log on to its public directories, and transfer files from your computer to that computer.

applet A program that was written in Java programming language that can be included in a HTML page, much like including a graphic image.

archie An Internet tool for finding specific files in publicly accessible archives.

ASAP WebShow Lets users view, download, and print graphically rich reports and presentations from the Web that were created by Software Publishing Company's WordPower software.

audio A multimedia element that can consist of human speech, music, or sound effects.

audio file format (SND, AU) An 8-bit monaural sound file format widely used on UNIX workstations, including NeXt and Sun machines for storing wave sounds. The format uses an advanced storage technique that enables 14-bit sounds to be stored in only 8 bits of data.

Autodesk flick format (FLI) Animation files that are created with the Autodesk Animator software package.

BinHex A method of encoding binary files so that they contain ASCII characters that can be easily transferred to other computers on the Internet.

bitmapped graphic A graphic image composed of a series of pixels.

browser A program that lets users view pages on the World Wide Web.

central processing unit (CPU) A computer's internal storage, processing, and control circuitry.

channel The basic unit of discussion on some computer conferencing programs.

chat Interactive, online conversations on the Internet in which participants key messages into their computers or terminals.

clip art A collection of graphic files stored on diskette or CD-ROMs that can be used in word processing, desktop publishing, or other application programs. Clip art is available in various graphic file formats.

codec A type of compressed file used in multimedia in which graphic files are compressed for efficient transmission or storage and decompressed for playback purposes.

computer conferenceing An electronic means of sending, viewing, and sharing real-time communications in areas of common interest via one or more of the following media formats: text, graphics, audio, video, and/or animation.

converter A type of HTML software utility program that converts an existing file type to HTML. Standalone converters work independently of other programs; add-on converters or templates are used in conjunction with an existing software application program, such as Microsoft Word or WordPerfect.

CU-SeeMe A videoconferencing system developed by Cornell University that is available via the Internet.

default The presetting that a program displays after it is installed.

editor A multiple-feature HTML software utility program, such as HTML Writer or HTML Pro.

electronic mail (e-mail) The most used of the Internet tools, allows users with an Internet connection who have been assigned a user identification (i.e. jsmith@psu.edu) to send text messages to others connected to the Internet.

external image Graphic image that must be downloaded separately and usually accessed through a link.

file transfer protocol (FTP) An Internet standard for the exchange of files.

gopher A UNIX-based system containing menus that helps you find files, programs, definitions, and other resources.

graphic image A multimedia element that can consist of a drawing, or static photograph, demonstrating no movement.

graphical image format (GIF) An 8-bit, bitmapped, 256-color format and works well with graphic lines or separator bars; icons, buttons, or bullets for lists; charts, tables, or graphics you want to place with text; small illustrations; or thumbnails.

handle An electronic pseudonym intended to conceal the user's true identity.

helper application (helper) An external application program that Netscape or another browser uses in order to retrieve a file with a format that Netscape or another browser itself cannot read.

homepage A document intended to serve as an initial point of entry in the World Wide Web.

hypermedia A method of linking to other connected servers and items on a server in which users click on text or other objects to link to other objects, text, pages, or on a particular place on a page on the World Wide Web.

hypertext A method of linking to other connected servers and items on a server in which users click on text to link to other objects, text, pages, or on a particular place on a page on the World Wide Web.

hypertext markup language (HTML) Provides a system for marking up text documents and a way to integrate multimedia and the use of hyperlinks.

hypertext transport protocol (HTTP) The Internet standard that supports the exchange of information on the World Wide Web.

inline image Graphic image that can be displayed directly on a Web page.

instancing One of the virtual reality modeling language (VRML) elements that refers to being able to use an object again, once it is defined.

integrated services digital network (ISDN) A digital telephone service that uses digital, rather than analog, signals. Transmission speeds of between 56 KB to 128 KB are possible.

Internet A network of computer systems that are interconnected in about 130 countries.

Internet relay chat (IRC) The most used UNIX-based computer conferencing program.

Java An application that provides a method to incorporate animation and sound into Web pages.

JavaScript The Java scripting language in which Java is used to accomplish a specific automated task in a Web document.

Joint Photographic Experts Group Format (JPEG) An 8- or 24-bit graphic image file format that takes advantage of human perceptual characteristics and does not deal with less essential information.

link Connection between two files or data items that allows a change in one to be reflected by a change in the other.

lurker A user who is connected to a CU-SeeMe videoconferencing reflector who is not sending video.

Lynx A popular nongraphical World Wide Web browser.

Macintosh audio information file format (AIFF) An 8-bit monaural sound file format developed by Apple Computer for storing digitized wave sounds. The format is also widely found on Silicon Graphics workstations and on the Internet.

Macromedia Shockwave A way of viewing interactive multimedia and high-impact graphics on the Web, installed in Netscape and other browsers as a plug-in.

Microsoft Windows audio format (WAV) A sound file format developed by Microsoft and IBM that calls for both 8-bit and 16-bit storage formats in both monaural and stereo. Most of the WAV sounds encountered on the Internet are 8-bit monaural sounds.

modem A computer peripheral device that connects a computer to a data transmission line. Most modems for personal use transfer data at speeds ranging from 1,200 bps to 33.6 Kbps.

Mosaic A popular graphical World Wide Web browser.

motherboard A large circuit board that contains the central processing unit (CPU) and other components.

Moving Picture Experts Group Digital-video Standard format (MPG or **MPEG)** A video file format that has become the de facto standard for digital video transmitted on the Web.

multimedia The integration of at least two media—text, photos, graphics, sound, music, animation, and full-motion video.

Musical Instrument Digital Interface format (MIDI, MID) A sound file format that serves as a standard communications protocol for the exchange of information between computers and music synthesizers.

netiquette Network etiquette representing rules that govern acceptable social behavior on the Internet.

Netscape Navigator A popular graphical World Wide Web browser.

packet-switched network A type of network in which information gets broken into little packets; these packets move independently of each other until they reach the destination.

plug-in module (plug-in) Software program that extends the capabilities of Netscape and other HTML browsers. Plug-ins can have one of three modes of operation: embedded, full-screen, or hidden.

point-to-point protocol (PPP) Allows a computer to use the TCP/IP (Internet) protocols with a standard telephone line and a high-speed modem.

portable document format (PDF) The file format created with Adobe Acrobat.

protocol The set of rules or conventions by which two machines talk to each other.

Quicktime Cross-Platform video file format (MOV) A video file format that was originally devised for use with the Macintosh computer.

raster-based See **bitmapped**.

real-time audio player A browser helper application that enables users to listen to live and rebroadcast audio on the Internet without needing to download, store, and play a sound file. Popular players include RealAudio from Progressive Networks and TrueSpeech from the DSP Group.

remote log on See **telnet**.

resource tree See **subject tree**.

scanner A hardware device used to convert digitized images into GIF or JPEG file format for use in a Web page.

search engine (spider) A program that searches the Internet, attempting to locate publicly accessible resources.

serial line IP (SLIP) Allows a computer to use the Internet protocols with a standard telephone line and a high-speed modem.

server The computer on which server software runs.

server software Software that allows a computer to offer a service to another computer.

service provider An organization that provides connection service to the Internet.

shell connection A dial-up service Internet connection in which the subscriber accesses the account with terminal emulation software.

spider See **search engine**.

subject tree In the World Wide Web, a guide to the Web that organizes Web sites by subject (i.e., Yahoo).

tag In HTML, an instruction to the Web browser about how to display text and other elements.

talking head A video technique in which an onscreen human image interacts with users.

telephony computer conferencing A type of computer conferencing in which participants are able to communicate via audio, similar to using a standard telephone, over the Internet. Programs such as Internet Phone or CoolTalk are needed.

telnet Allows users to log on to any computer connected to the Internet from another computer which also has an Internet connection.

thumbnail A miniature graphical representation of a document page.

transmission control protocol/Internet protocol (TCP/IP) The type of protocol used by most servers connected to the Internet.

uniform resource locator (URL) A string of characters that precisely identifies an Internet resource's type and location (i.e., http://www.microsoft.com).

UNIX An operating system used on a wide variety of computers that are connected to the Internet.

veronica A search tool in gopher (see gopher) that scans a database of gopher directory titles and resources and generates a new gopher menu containing the results of the search.

video A multimedia visual element that can consist of full motion video or animation.

video capture A method of acquiring an image file using a video capture card installed in one of the expansion slots of an IBM or IBM-compatible computer and software that lets users capture images with an attached video camera.

video capture card An adapter that plugs into a computer's expansion bus that enables you to control a video camera or videocassette recorder (VCR) and manipulate its output.

videoconferencing A full-motion, two-way, video/audio system that permits two or more people in different locations to communicate with each other.

virtual reality modeling language (VRML) A language for describing multiparticipant interactive simulations in which the virtual worlds are networked through the World Wide Web.

virtual reality (virtual world) A created environment using a virtual reality modeling tool, within which users operate and interact.

Web See **World Wide Web**.

Web browser See **browser.**

Web page In the World Wide Web, a set of related documents that make up a hypertext presentation.

Web server A program on a computer connected to the World Wide Web that accepts requests for information according to hypertext transport protocol (HTTP).

Web site In the World Wide Web (WWW), a computer system that runs a Web server and has been set up for publishing documents on the Web.

WinSock A common term for Windows Sockets, a set of specifications that programmers must use to write TCP/IP software for Windows.

World Wide Web (WWW or Web) A global hypermedia system that uses the Internet as its transport mechanism; a network of networks.

Yahoo A popular subject tree for the World Wide Web.

SELECTED BIBLIOGRAPHY

Books

Deep, John and Holfelder, Peter. *Designing Interactive Documents with Adobe Acrobat Pro.* New York, NY: John Wiley, 1996.

Eager, Bill. *Using the Internet.* Indianapolis, IN: Que Corporation, 1994.

Fox, David and Downing, Troy. *HTML Web Publisher's Construction Kit.* Corte Madera, CA: Waite Group Press, 1995.

Gates, Bill. *The Road Ahead.* New York, NY: Penguin Books, Inc., 1995.

Grant, Kenneth D. and Schwaderer, W. David. *Adobe Acrobat Handbook.* Carmel, IN: Sams Publishing, 1993.

Heslop, Brent and Budnick, Larry. *HTML Publishing on the Internet for Windows.* Chapel Hill, NC: Ventana Press, 1995.

Hoffman, Paul E. *Netscape and the World Wide Web for Dummies.* Foster City, CA: IDG Books Worldwide, 1995.

Korolenko, Michael D. *Writing for Multimedia: A Guide and Sourcebook for the Digital Writer.* Belmont, CA: Wadsworth, 1997.

Magee, Sasha and Rabinowitz, Sasha. *Shockwave for Director User's Guide.* Indianapolis, IN: New Riders Publishing, 1996.

Mara, Mary Jane. *Web Head: A Mac Guide to the World Wide Web.* Berkeley, CA: Peachpit Press, 1995.

Perry, Paul J. *World Wide Web Secrets.* Foster City, CA: IDG Books Worldwide, 1995.

Pesce, Mark. *VRML: Flying Through the Web.* Indianapolis, IN: New Riders Publishing, 1996.

Pfaffenberger, Bryan with Will, David. *Que's Computer & Internet Dictionary.* Indianapolis, IN: Que Corporation, 1995.

Pinnheiro, Edwin. *Introduction to Multimedia Featuring Windows Applications.* Belmont, CA: Wadsworth, 1996.

Sattler, Michael. *Internet TV with CU-SeeMe.* Indianapolis, IN: Sams.net Publishing, 1995.

Tittel, Ed and Gaither, Mark. *Mecklermedia's Official Internet World 60 Minute Guide to Java.* Foster City, CA: IDG Books Worldwide, 1995.

Wilson, Stephen. *World Wide Web Design Guide.* Indianapolis, IN: Hayden Books, 1995.

Unpublished Writings

Azami, Rodney. "Multimedia Presentation Levels." Chart presented by College of Business Administration, California State Polytechnic University, Pomona, CA, at the Continuous Improvement Seminar, American Association of Collegiate Schools of Business, October 1995.

Wilson, Jamie. "Purchasing a Multimedia Computer." Paper presented at the National Business Education Association Convention, San Francisco, CA, April 1995.

INDEX